Coaxial AC Bridges

Coaxial AC Bridges

B P Kibble and G H Rayner

Adam Hilger Ltd, Bristol

British Library Cataloguing in Publication Data
Kibble, B. P.
  Coaxial AC bridges.
  1. Electric circuits—Alternating current
  2. Bridge Circuits
  I. Title   II. Rayner, G. H.
  621.319′13   TK454

  ISBN 0-85274-389-0

Consultant Editor: **Mr A E Bailey**

Published by Adam Hilger Ltd, Techno House, Redcliffe Way, Bristol BS1 6NX.

Phototypeset by Quadraset Ltd, Radstock and printed in Great Britain by J W Arrowsmith Ltd, Bristol.

Contents

Preface

There has been a radical change over the past two decades in the experimental bridge network techniques practised by national metrological laboratories to compare electrical impedances. This has come about chiefly because of the work of R D Cutkosky and A M Thompson, who have shown great kindness in instructing us personally. Unfortunately, a more general acceptance and use of these techniques is hindered because they are described in a somewhat piecemeal fashion in articles in various journals. We have therefore attempted in this book to provide a more systematic treatment, not just for the use of those seeking to practise electrical metrology, but for electrical engineers in general, because the basic principles of conductor-pair networks can be applied elsewhere, particularly in the design and use of electronic measuring instruments.

The extensive existing literature on AC bridges includes a very complete book written by Hague and revised by Foord. Our book is not in any sense to be seen as replacing this work; we aim at a self-contained exposition of new techniques which can be applied to the traditional bridge networks if desired. The approach is mainly practical, with attention paid to constructional principles and techniques. We strongly urge any reader who has difficulty with the concepts described (particularly in chapter 2), and who has the facilities, to set up some of the simpler networks and devices for himself and to experiment with them. Even the simpler comparison networks when constructed with suitable components are capable of an accuracy of a part in a million or better; accuracies approaching a part in 10^9 are achieved in national metrological laboratories.

There is a beauty somewhat akin to a musical or artistic experience in the design of a good conductor-pair bridge arising from the completeness and inevitability of development from abstract propositions and definitions, and this has provided the strongest appeal of the subject for us. We hope that some readers will find the same satisfaction.

B P Kibble
G H Rayner
National Physical Laboratory
Teddington, 1983

1 Introduction

1.1 Improvements in defining standards and in bridge networks_____

Early comparisons of the values of impedances were limited to resistance measurement because only direct current sources and detectors were available. A little later some form of commutator was used to provide intermittent or reversed current, and this allowed impedances which are only defined for the passage of time-varying currents, namely capacitors and mutual and self-inductors, to be compared. However, bridge design continued to revolve around DC techniques for which the spatial disposition of conductors, impedances, detectors and sources is immaterial, and the basic standard of impedance was a DC resistor.

Even after sinusoidal sources and detectors responding to a single frequency became available and some of the advantages of AC bridge comparisons became apparent, the past history of the subject lingered on in the form of single-conductor networks used for AC bridges, but the recent advent of the calculable mutual inductor and the even more accurate calculable capacitor has necessitated the development of better techniques. (A particular form of calculable capacitor is capable of yielding an impedance unit, the farad, from one length measurement, the defined magnetic permeability of free space, and the defined velocity of light with an accuracy of a few parts in 10^8.)

Electrical impedances store or dissipate the energy of magnetic or electric fields and currents, and so for a reproducible well defined standard of impedance the location and strength of the associated fields must be defined by fixing the boundary of the space they occupy. A balanced bridge network relates the energy flow through two or more sets of impedance standards in order to compare their values accurately; to do this the standards need to be joined by conductors to form a network. Therefore the magnetic and electric fields surrounding such conductors must also be controlled and accounted for, as they also constitute an energy flow. If the impedance standards have exposed terminals and single wire conductors are used to connect to each terminal, fields surround both conductors and standards and the strength

1

and distribution of these fields is altered by changes in relative position of the standards, conductors and external objects, including the experimenter. The use of such networks inevitably leads to imprecise relationships between the values of the standards because the stray fields associated with the standards and networks can be altered by changes in their environment.

The main object of this book is to show how, by using coaxial conductors which carry equal and opposite currents in their central and concentric outer conductors together with electrical and magnetic screens surrounding components and standards, a bridge network can be well defined in the above sense. Then *certain and meaningful* measurements can be made, with accuracies of 1 in 10^9 if required.

This new approach may be viewed simply as the application of the coaxial techniques common in radio-frequency work to low frequency bridges, combined with more careful and precise definition of what is meant by the impedance of a device used as a standard. The ideas have been developed by many people over a period of years, but principally by R D Cutkosky of the National Bureau of Standards, USA and A M Thompson of the National Measurement Laboratory, Australia.

One way of looking at the bridges we describe is to see them as two superposed networks. The first of these consists of straightforward meshes of components and the interconnecting wires between them. The second network comprises the screens of the components and the outer, coaxial screen of the connecting cables. The configurations of the two networks are identical and by providing every mesh with an equalising device, the current in the outer screen is constrained to be equal in magnitude and of opposite sign to the current in the components and central conductors. The current in any cable as a whole is zero and no external magnetic field is created. The second network of screens and cable outer conductors has a low impedance and it is all at nearly the same potential, so that there is no significant external electric field. This construction has the further important advantage that such networks do not respond to fields from external sources, whether at the source frequency or an incoherent interfering frequency.

The standards compared by these networks of conductor-pairs must be provided with terminal-pairs. The physical form of a terminal-pair will be a coaxial connector, with the inner contact being the terminal connected to the component, and the outer being the terminal connected to its enclosing conducting case. A coaxial terminal-pair is often called a port. Just as with low-value resistors used in DC work, where it is usually unsatisfactory for a standard to be provided with only two terminals because of varying contact resistance, and so separate current and potential terminals are provided to make a four terminal standard, a standard provided with four terminal-pairs is required for AC work of the highest accuracy. Moreover, the electrical conditions for which its defined value is obtained need to be specified precisely. But just as a two terminal construction may be adequate for DC resistors of high value, in AC work a simpler two terminal-pair standard may often be adequate.

Unfortunately, the majority of low-frequency reactance standards in laboratories and being manufactured are not provided with suitable coaxial connections, but it is usually a simple matter to replace the existing terminals and where necessary to provide a new outer conducting container.

It is very much the purpose of this book to encourage coaxial techniques within the formalism of conductor-pair networks, to show that these techniques have the great advantage of obtaining precise and certain results and to show a way to implement each concept with constructable apparatus.

The bridge networks with which we are concerned will therefore consist of standards provided with, and defined in terms of, terminal-pair coaxial terminations and will be connected together with conductor-pair coaxial cables. Balancing the currents in the inner and outer of cables so that they are equal and opposite will ensure that there is negligible external magnetic field. Theoretically, zero magnetic field is obtained only for a straight infinite coaxial balanced cable, but in practice short lengths and quite severe curvature can be tolerated. As mentioned above, the external electric field is also negligible.

In general, the techniques we describe are appropriate for frequencies from 10 Hz to at least 1 MHz. Highest accuracy is attained for frequencies from about 400 Hz to 2 kHz.

At first sight, passive conductor-pair networks seem unrelated to the design of electronic circuits and equipment, but, because of its emphasis on a complete understanding of circuits and their interactions, AC bridge work is a valuable subject to study for anyone concerned with any kind of electrical circuit. Unexpected interactions between apparently unrelated parts of a circuit are a frequent source of frustration, if not outright failure. The equalised conductor-pair concept is also a powerful design technique for electronic circuits, although somewhat different means such as the use of differential input amplifiers are employed to achieve equalisation or minimisation of loop currents. The role of power supply leads and their associated current return conductor ('earth') to and from a packaged circuit needs careful consideration in relation to the network as a whole.

1.2 Ratio devices

Improved ratio devices are now available which enable more accurate and convenient bridge networks to be made. These ratio devices are current or voltage transformers, constructed so that nearly all the magnetic flux threading one winding also threads another. This is accomplished by an appropriate winding geometry, usually on the surface of a toroid, and by providing a low-reluctance path for the flux with a magnetic material of high permeability. Then the ratio of EMFs induced, or currents flowing, in the two windings can be simply the ratio of the numbers of turns of the two windings to within a part in a million or better. The small departures from perfection are fairly constant and can be measured and allowed for in high accuracy work.

Furthermore, transformers can also fulfil the function of isolation, that is, they can relate currents or potentials in two sub-networks of a bridge without allowing any net current to flow between them, and as a consequence of this property the potentials produced by separate windings can be strictly added in any order. Attention must be paid, however, to eliminating capacitative currents between the windings by proper electrical screening techniques.

1.3 Impedance standards

We have seen in §1.1 the need to construct standards with no significant electric or magnetic fields outside their containers. They must be provided with two or more pairs of terminals which should take the form of coaxial sockets; the inner connection is one terminal and the outer connection the other terminal of a pair. The value of the standard must be strictly defined as conditions of potential difference between terminal-pairs and of a current flow into one terminal and an equal and opposite current out of the other terminal of a terminal-pair. The simplest definition of this kind is that of a two terminal-pair standard whose impedance is defined as the ratio of the potential difference between one specified terminal-pair to the current flowing at the other terminal-pair, when there is zero potential difference between the terminals of this terminal-pair. The specified defining conditions may be departed from provided this departure can be precisely related to the definition. For work of the highest accuracy the conditions at a single terminal-pair may not be sufficiently reproducible, chiefly because of variations in contact resistance with the mating coaxial connectors, and then it is necessary to remove the defined points back into the interior of the device and to provide two additional terminal-pair connections. For example, a voltage source may be connected to one terminal-pair and the other terminal-pair used to sense the actual voltage at points within the device. Thus an impedance requires four terminal-pairs in order to apply this principle, in an analogous manner to the four terminals of the familiar four terminal DC resistor. Other definitions of standards are used, but they can be viewed as degenerate examples of the two cases above.

A good deal of space will be devoted in chapter 2 to further careful treatment of the electrical definition of the values of standards; we must emphasise that this is most certainly essential. Once the defining conditions of standards whose values are to be compared are settled on, the design of proper comparison networks will be seen to follow naturally. Neglecting or skimping the definition stage will, as in any other accurate measurement, inevitably lead to failure.

Standards can be represented by lumped inductances, capacitances and resistances connected in a network; that is to say the behaviour of the actual device cannot be distinguished at the frequencies with which we are presently concerned from the behaviour of such an equivalent network by making any measurement at the terminals of the device.

Another minor advance in technique is that in most cases, the variable or decade switched impedances used in older bridge technology can be replaced with advantage by components of fixed value, with currents driven through them by a variable voltage from a multi-decade ratio transformer.

1.4 The concept of 'ground' or 'earth' connections_____

These concepts, of historical origin, have caused much confusion and misery to practitioners of AC measurements, and have moreover brought the whole subject somewhat into disrepute, mainly because of the woolly manner in which the concepts are frequently invoked. We suggest as an alternative that one point of a network be designated a potential reference point to which all other potentials of the network are related, and which is therefore usually given the value of zero potential. It will usually be at about the same potential as the 'earth' or surroundings of the apparatus—walls, water pipes, operators etc—bearing in mind that their potentials are rather uncertain.

In addition, it must be realised that a current flows in a complete circuit under a continuous fall in potential from its source back to its source; it does not just go to 'earth' or 'ground' and disappear; neither does the extended 'ground' of most electronic instruments constitute an equipotential conductor, if currents flow through it.

1.5 Formal representation of circuit diagrams and components_____

We have found a formal representation of the coaxial cables and components used in the bridges discussed in this book to be useful, and we believe the reader will also. Initially, however, it may help to draw the coaxial cables and components in a more pictorial fashion, and to envisage the actual physical layout. We here introduce the elements of the formal representation (which is adapted from the conventions used in the papers of Cutkosky and Thompson). Some of the conventions used, particularly for lines representing conductors crossing, do not conform to modern practice, but we find that they clarify the representation of a conductor-pair network.

The formal representations of particular devices are introduced with the description of the devices as they occur throughout the book.

The bridge techniques, as we have seen, are based on the use of conductor-pair networks where, usually, one conductor is the inner conductor of a coaxial cable with the other the surrounding outer conductor. Such a cable could be drawn pictorially as in figure 1.1, but we prefer to represent it formally as in figure 1.2 with the inner conductor as a fine line and the outer conductor as a thicker line alongside it (to suggest its low impedance per unit length).

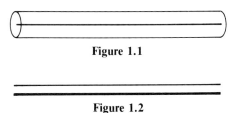

Figure 1.1

Figure 1.2

The cables terminate at a junction with other cables or components, the connection being accomplished by mating coaxial plugs and sockets. This interface (between two cables) could be drawn pictorially as in figure 1.3, but we prefer to represent it by small circles on both inner and outer conductors, as in figure 1.4, to emphasise the concept of the mating planes of both inner and outer connectors forming *terminals* by analogy with the screw terminals used as net nodes in older bridge techniques. It is a junction of this kind which we call a terminal-pair. It is important that the mating surfaces of the connectors give a reliable low-resistance connection. Since the frequencies we are concerned with are low, impedance matching of cables and connectors is unnecessary.

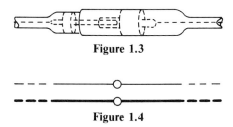

Figure 1.3

Figure 1.4

The symbol —○ therefore represents a terminal or junction point to which external connections can be made; we will use —●— to represent a junction point internal to a device, to which additional connections cannot be made.

An immittance represented by figure 1.5 may be either
(i) an impedance (Z) having reactive ($j\omega X$) and resistive (R) components,
(ii) an admittance (Y) having susceptive ($j\omega B$) and conductive (G) components.

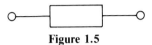

Figure 1.5

Where the details of a general network of immittances are not of immediate concern to us, the network will be represented as in figure 1.6 as a cloud with one or more terminal-pairs as appropriate for making connections to the network.

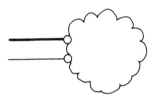

Figure 1.6

The familiar symbols of electrical circuit theory are used for resistors, capacitors and inductors which form the lumped representation of actual physical components; they will in general be surrounded by a screening and conducting case to which the outers of the coaxial cables are joined, and so a typical two terminal-pair component can be represented as in figure 1.7.

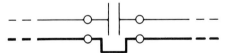

Figure 1.7 *Diagrammatic representation of a two terminal-pair component, with attached coaxial cables.*

There is perhaps an implication that the component and its surrounding shielding box also have coaxial geometry; this would be an ideal situation, but the limitations of actual practical manufacture do not often permit it. The external magnetic fields generated by a non-coaxial geometry can be eliminated if necessary by a magnetic shield.

Conducting surfaces (electrical screens) will be represented as broken lines. Where single conductors are necessary, they will be represented as curved flowing lines with a loop to suggest their unpaired and unbalanced nature in that currents in them give rise to surrounding magnetic fields.

Transformers deserve special attention, both because they are principal components in all the networks described, and because their constructional topology is crucial to their accuracy. All of the transformers we shall be considering have a toroidal core made from a material of high permeability. A toroidal form is preferred because of its circular symmetry so that if it is excited by a uniform winding, very little magnetic field exists external to the toroid. Some provision should be made for cancelling the magnetic effect of the advancing winding, which is equivalent to a single annular turn around the toroid (figure 1.8).

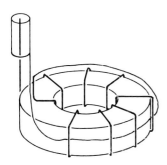

Figure 1.8

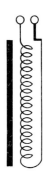

Figure 1.9 *Diagrammatic representation of the wound toroid of figure 1.8.*

The windings of the transformer have to thread through the central space as each turn is made. To show this diagrammatically, as in figure 1.9, the following convention will be used. First, on the left, is the representation of the core as a line; next to this is one winding and if this is completed by making a reverse annular turn parallel to the toroid, its return is shown as a straight line parallel to the core. The ends of the winding terminate on the inner and outer of a coaxial socket, or terminal-pair. In the more elaborate example drawn in figure 1.10, there is an electrical screen between the windings, shown as a broken line. This is followed by a second winding and a final electrical screen is shown to the right, it thus being understood that a given winding or screen threads all windings or cores drawn to the left of it.

Other specialised or more elaborate transformers will be depicted similarly —the order in which the component parts are shown in the diagram follows the order in which a transformer is assembled, starting with the core and working systematically outwards through the windings and shields.

Two windings applied simultaneously side by side or, for example, as a twisted pair of conductors, are drawn one above the other as in figure 1.11.

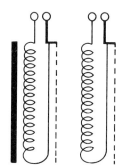

Figure 1.10 *A transformer core with two windings,*
each having an overlaying electrical screen.

Sometimes it is convenient to draw an assembly from right to left, or from top to bottom, or from bottom to top instead of from left to right as above, and to leave the context to clarify the intended construction.

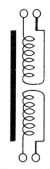

Figure 1.11

We will always represent real bridge sources as voltage sources (that is, sources of negligible series internal impedance), and add the lumped representation of the internal impedance of an actual source, if of any importance to the network in which it occurs. This is not usually the case, and so the internal impedance will then be omitted. Thus an oscillator and its output coaxial termination will be drawn as in figure 1.12 if its output impedance is small so that it approximates a zero impedance voltage source and as in figure 1.13 if it is the dual of this, namely as an approximately infinite impedance current source.

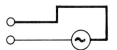

Figure 1.12 *A voltage source with*
a coaxial output termination.

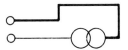

Figure 1.13 *A current source with*
a coaxial output termination.

We will initially ignore the extra complication brought about by connections to the mains supply which are either direct or indirect via the interwinding capacitance of supply transformers in sources or detectors. Isolation can actually be accomplished reasonably well by proper design of a screened mains transformer. We discuss this whole aspect of isolation later, but for the present it will be convenient to assume its accomplishment. For example, we can at present assume that all active components (generators, detectors, amplifiers, etc) are powered by batteries wholly contained within the instrument's conducting external screening case.

Thus the source and detector for a bridge network will consist of an isolated box provided with an input coaxial connector. A detector is represented as in figure 1.14.

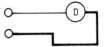

Figure 1.14 *A detector with a coaxial input port.*

1.6 Achieving equal and opposite currents in the inner and outer conductors of coaxial cables

Equalisation of currents in networks at audio frequencies can be achieved by a simple technique whereby somewhere around each mesh of the conductor-pair network a cable is threaded a few times through a high permeability toroidal core.

Consider one mesh, drawn with its equivalent circuit in figure 1.15. We will find that in the circuit of the inner conductor there will be a source ε and a relatively high impedance Z. In contrast, the circuit of the outer conductor will have only a low impedance z. We can redraw the mesh to show that the core and cable acts as a 1:1 transformer, the inner conductor being the

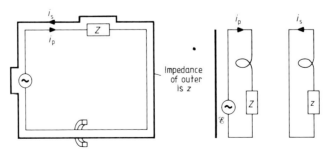

Figure 1.15 *One mesh of a coaxial network threaded through a core and its equivalent circuit.*

primary and the outer a low impedance secondary. The effectiveness of the device can be enhanced by winding the cable through the core n times.

Let Φ_P and Φ_S be the magnetic fluxes in the core caused by the primary and secondary currents i_P and i_S respectively. The total flux in the core,

$$\Phi = \Phi_P - \Phi_S = n(i_P - i_S).$$

Then $n\,d\Phi/dt = \mathcal{E} - i_P Z = i_S z = n^2(di_P/dt - di_S/dt) = j\omega L(i_P - i_S)$ for a sinusoidal current. Hence

$$i_P = i_S + i_S z/j\omega L \text{ and if } |z| \ll |j\omega L|$$

$$i_P \simeq i_S \tag{1.1}$$

and

$$\mathcal{E} \simeq i_P(Z + z) \tag{1.2}$$

Equation (1.1) shows that the equality of currents has been achieved, to a good approximation. Typical values of ωL and z would be 100 and 0.1 ohms respectively.

Equation (1.2) shows that the apparent impedance of the primary circuit has been increased by z. In general, in equalised bridge networks, the measured impedance of a two terminal-pair standard is that of the impedance in the circuit of the inner conductor increased by the corresponding small impedance of the outer conductor. We will return to this point in §2.5. In §4.4.7 we will discuss how, in practice, the effectiveness of a particular current equaliser employed in a network can be tested.

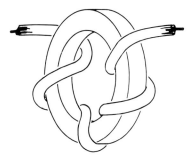

Figure 1.16 *A current equaliser.*

A current equaliser, shown pictorially in figure 1.16, will be represented diagrammatically as in figure 1.17 by a symbol for a transformer core drawn alongside the outer conductor.

Figure 1.17 *The diagrammatic representation of a current equaliser.*

1.7 A simple example of a conductor-pair bridge_____

Figure 1.18(*a*) shows some of the elements so far discussed combined in a simple capacitance bridge, with an alternative pictorial view in figure 1.18(*b*) of what is symbolically represented.

In some networks we will require an isolated source, that is, one whose potential with respect to a reference point can be made to take any value. It will be drawn as shown in figure 1.19, and can be practically realised by means of a conventional source as above in conjunction with an isolating transformer, as in figure 1.20.

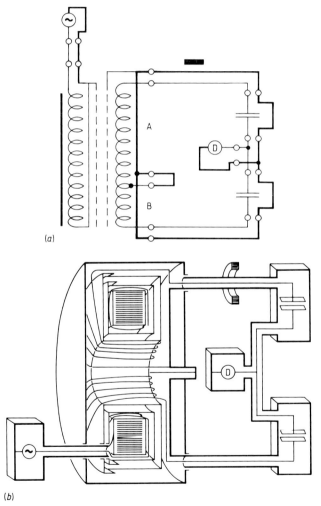

(*a*)

(*b*)

Figure 1.18 *(a) A simple coaxial capacitance comparison bridge. (b) A pictorial section view of the complete capacitance bridge of (a).*

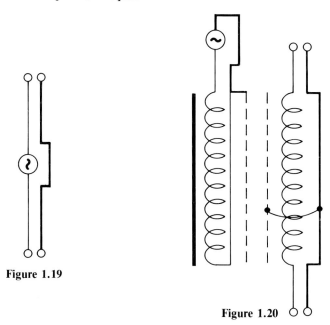

Figure 1.19

Figure 1.20

1.8 The relationship between coaxial conductor-pair techniques and the older bridge literature _____

Bridges are drawn as single conductor networks both in the older bridge literature and at present when it is desired to illustrate their working principle. This representation can be converted to the conductor-pair bridges we wish to discuss by taking one point, the reference point or 'earth', and topologically stretching and extruding it as tubes over all the single conductors, components, detectors and sources. When current equalisers are added to each separately identifiable mesh, the result is a conductor-pair network. The employment of coaxial conductor-pair techniques in the familiar four-arm bridge leads to large shunt impedances from either end of every component to the coaxial outer conductor. The means of eliminating the effect of these shunt impedances from the balance condition for a bridge will provide the first example of the auxiliary balances needed in high accuracy conductor-pair networks. Figure 1.21 shows a single conductor four-arm bridge.

This is redrawn below in figure 1.22 as a four-arm bridge with complete screening. Variable two terminal impedances which are in addition to the shunt impedances already present have been added to the four junction nodes of the components. In order to appreciate the purpose of these components, it is convenient to redraw the network of the outer conductors as a simplified line, as in figure 1.23.

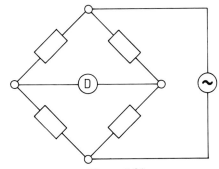

Figure 1.21

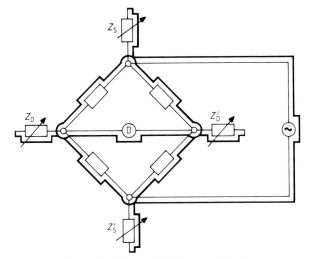

Figure 1.22 *A coaxial four-arm bridge.*

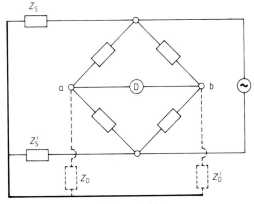

Figure 1.23

The main balance having been made by nulling the detector D, the bridge nodes a and b can be brought to the same potential as the outer conductors by adjusting Z_S, Z_S'. As there is then no potential difference across them, it is evident that the shunt impedances Z_D, Z_D' do not then affect the balance condition, whatever their value. This auxiliary adjustment is termed a Wagner balance in bridge literature.

If we consider the interchange of the source and detector in the network, which from the principle of conjugacy does not affect the balance condition, it can be seen that the addition of adjustable impedances Z_D, Z_D' is also

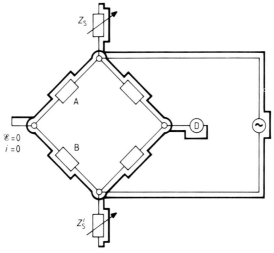

Figure 1.24

desirable. Reverting to the original source and detector positions, if the Wagner components are deliberately misset, a and b will not be at the same potential as the outer conductors, but they can be at the same potential as each other (that is, D can again be nulled and the bridge balance condition be restored) if the ratio of Z_D to Z_D' is adjusted. This operation may be termed the detector network auxiliary balance. It is evident that, when the previous adjustment of the Wagner components is restored, the product benefit of both auxiliary adjustments has been gained, so that, for example, adjustment of both with an accuracy of 1 in 10^3 is sufficient to ensure that the bridge balance condition is not affected by shunt impedances to more than 1 in 10^6. The bridge described in §1.7 is also a four-arm bridge, having a ratio transformer forming two of the arms. A required defining condition for such a transformer is that there should be no current flowing through the short on the tap, as indicated in the first diagram of §1.7 where the voltage ratio arms are A and B. In the simplified layout of figures 1.24 and 1.25 it can be seen that this condition is achieved by

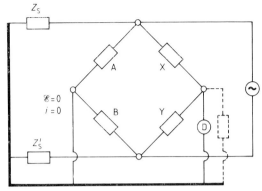

Figure 1.25

adjustment of the ratio of Z_S to Z_S'. If D has previously been nulled, shunt impedance at the node to which it is connected will merely affect its sensitivity, and not the bridge balance condition. It is a property of a properly constructed ratio transformer that the voltage ratio is hardly affected by shunt impedance loads at the output ports at the upper and lower bridge nodes, so the effect of these shunt impedances on the balance condition is often less than the required accuracy, and so no second auxiliary balance is required. A more formal discussion of the concepts of this section is given in §2.6 below.

Reference

Hague B (revised by T R Foord) 1971 *Alternating current bridge methods* 6th edn (London: Pitman)

2 General Principles

We have already emphasised the need to define completely the electrical conditions to be obtained at the terminals of a standard before designing a suitable measurement network or bridge to relate it to other standards. The defined conditions can rarely, if ever, be realised exactly, and so we need to consider whether the actual conditions attained are either equivalent to the ones sought, or are a sufficiently good approximation to them to achieve the desired accuracy. In this chapter we consider in detail firstly a number of possible terminal conditions which may be used to define a standard and secondly the effect of connecting coaxial cables to these terminals.

If the concepts presented in this chapter are new to the reader, a detailed study could be deferred until the application to the bridge networks described in chapter 6 is apparent.

2.1 The validity of lumped component representation_____

The flow of electrical energy driven by an alternating EMF through an actual electrical component can be represented by considering the component to be a network of some or all of the idealised components of inductance, L (in which the electrical energy is associated only with the magnetic field), of capacitance, C (in which electrical energy exists only in the electric field) and of resistance, R (in which the energy is dissipated entirely as heat and neither magnetic nor electric fields arise).

In an actual component, all three processes take place together within the device and in the surrounding space. Consider a coil of wire carrying a current. Each short section of the wire will be surrounded by a magnetic field so that it possesses inductance, will dissipate energy through its resistance and, as it is at a different potential from its neighbouring sections, there will be electric fields within and in the vicinity of the coil.

At sufficiently low frequencies these magnetic fields from successive sections will all have the same phase and the sum field can be treated as a single entity as if it arose separately from an ideal inductance. In a similar way, the effect of electric fields can be modelled as an ideal capacitance. The resistive

17

dissipation of the device can also be treated collectively as an ideal resistance in series with the inductance in the network of ideal components drawn in figure 2.1 which represents the actual behaviour of the device.

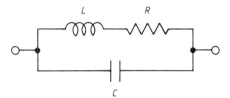

Figure 2.1 *An equivalent circuit for a coil.*

This simple model accurately represents the actual behaviour of the device up to frequencies higher than the self-resonant frequency $\omega_r = (LC)^{-1/2}$ of this representation because the self-resonant frequencies of the individual sections are even higher, and the electric and magnetic fields from each add to give the total field of the device in a way which still gives a constant mean phase throughout the volume of the device. Thus representing the actual behaviour of the device by lumped component parameters is valid provided the wavelength of the oscillatory magnetic and electric fields associated with the device is long compared with its dimensions.

If two devices are separated so that their electric and magnetic fields do not share a common region of space, their combined behaviour when connected together is represented by the connection of their respective ideal networks; this is not the case if their fields partially overlap. For example, two equal inductors L connected in series have a combined inductance $2L$ when well separated, but if they are brought together their combined value approaches one of the theoretical limits, $4L$ or 0, depending on the sense of connection. In general, therefore, it is the spatial distribution and interaction of the electric and magnetic fields of a device with its conductors which should be considered when deciding whether a particular lumped parameter representation is appropriate.

2.2 General principles applying to all impedance standards___

2.2.1 The physical definition of a standard
The value of a standard of *reactance* is defined by the energy stored in its electric field E or magnetic field H per unit of applied EMF or current flowing. In the cases of capacitance or inductance, the energy is $U = C\mathcal{E}^2/2$ or $Li^2/2$. These energies can be written as volume integrals in terms of the fields as $\int \epsilon\epsilon_0 E^2 \, \mathrm{d}V/2$ or $\int \mu\mu_0 H^2 \, \mathrm{d}V/2$. It therefore follows that any neighbouring

object which modifies these electric or magnetic fields will alter the value of the standard. A good standard will be so constructed and used that no appreciable modification can occur. The construction of the standard and the manner in which it is to be measured therefore need to be clearly laid down, and a satisfactory way of doing this could be termed the physical definition of a standard. Discussion of the ways of achieving an adequate physical definition and the extent to which available standards are satisfactory in this regard will be deferred until chapter 3.

A standard of resistance of value R should only dissipate energy at a rate $W = i^2R$; the geometry of its conductors should be such that the stored energy of its electric or magnetic fields, $(Li^2 + C\mathcal{E}^2)/2$, is small by comparison with the energy $(2\pi/\omega)i^2R$ dissipated per cycle.

2.2.2 The electrical definition of a standard

The electrical conductors of a measurement network are connected to terminal points on the standards. We will call the electrical definition of a standard the impedance defined as the ratio of the potential which exists between two designated terminal points to the current which flows through the same or two other terminal points. The admittance of a standard is the reciprocal of this ratio. Additional potential or current conditions may be imposed at the same or other terminal points to increase the precision of a definition, and certain kinds of definition are advantageous. We now consider them in a general, abstract way in preparation for the practical details of chapter 3.

It must be realised at the outset, however, that the electrical definition of a standard, or conditions which can be shown to be equivalent to it, must be strictly realised in a balanced bridge network if accurate results are to be obtained. For example, considering the familiar case of a four terminal DC resistor, the electrical definition is realised in practice in the potentiometric method of comparison in which the same current flows through two such resistors and the potential differences between their respective potential terminals are compared under conditions of zero current flow through these terminals. However, the same defined value is obtained by using Kelvin's double bridge described in §2.7, although in this network appreciable currents flow through the potential terminals. The reason for the equivalence of these conditions to those of the proper electrical definition will be discussed there.

2.2.3 Two terminal definition

The standard is provided with only two terminating points and its impedance Z is defined as the ratio of the potential between them to the current leaving one of them as shown in figure 2.2. Since there are only two terminal points and it is implicit in this definition that capacitance and conductance to the

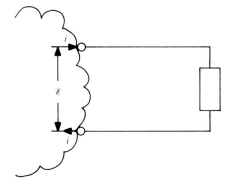

Figure 2.2 *A standard defined in a two terminal manner.*

surroundings is negligible, as otherwise the standard is ill defined, the current leaving the standard is equal at all times to the current entering.

This simple definition is electrically unsatisfactory, in that only an approximation to a terminal *point* can be obtained because actual terminals are objects of finite physical size having potential differences between different points on them. Moreover, uncertain potential differences arise through uncertain contact resistance, mutual inductance and capacitance between the measurement leads, the terminals and the conductors beyond them. Hence more elaborate defining conditions which overcome these problems are used for accurate standards.

2.2.4 Four terminal definition
This definition, illustrated in figure 2.3, may already be familiar as that used for accurate DC resistors because it eliminates any uncertainty due to indeterminate terminal resistances. The standard is provided with two extra terminals between which the potential difference is defined; these are connected, by conductors whose dimensions are long compared with their cross section, to internal points on the conductors going to the terminals which would have been concerned with a two terminal definition. We will call these internal points the *internal defining points*. Conditions of current flow and potential distribution at these points must not be affected appreciably by the variable conditions of flow and potential distribution at

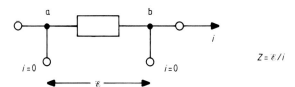

Figure 2.3 *A four terminal definition of a standard.*

the actual terminals if the electrical definition is to be precise. The usual arrangement of the four terminals is only suitable for DC or for low-frequency measurements and low-value impedances, as considerable electric and magnetic fields emanate from them and the conductors going to them.

2.2.5 *Four terminal coaxial definition*
This unfamiliar definition, illustrated in figure 2.4, is of great usefulness in resistance measurement. It is formally identical to that of the previous section, but the physical arrangement of the terminals is altered to two terminal-pairs and hence connections can be made to the standard in a proper equalised coaxial manner. It will be seen that the comparatively large and uncertain capacitance between inner and outer of the coaxial cables shunts the standard, and this is what makes the definition suitable only for resistive components.

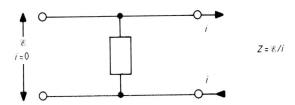

Figure 2.4 *A four terminal coaxial definition of a standard.*

2.2.6 *Two terminal-pair definition*
An alternative definition is more appropriate for reactance standards. The impedance is again defined as the ratio of the potential between two terminals to the current out of one of the other two terminals. When used in a properly equalised conductor-pair network there will be an equal current flowing into the fourth terminal and the additional condition is imposed that there shall be no potential between the terminals at which the current flow is defined.

It will be seen, in figure 2.5, that the currents into and out of the left-hand terminal-pair must also be equal since no net current leaves from the pair of right-hand terminals, but in general this current will not equal i, and its actual value plays no part whatever in the definition. Since the terminals between which the potential is defined on both the right-hand side and the left-hand side carry current, the finite size and impedance of an actual terminal will cause poor definition, but if the terminals and cables going to them are coaxial and the go and return currents equalised, there will be no significant external electric or magnetic fields to cause any additional lack of definition. There will in general be admittances Y_{mn} internal to the device

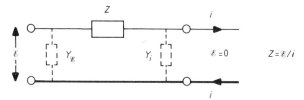

Figure 2.5 *A two terminal-pair definition.*

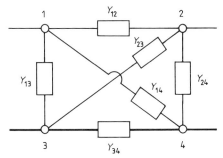

Figure 2.6 *The internal admittances of a two terminal-pair standard.*

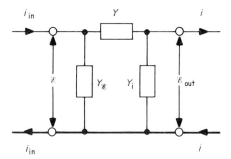

Figure 2.7 *A simpler representation of the internal admittances.*

(figure 2.6) which may be represented more simply as in figure 2.7 as just three lumped admittances, a *direct* admittance $Y = 1/Z$ and two shunt admittances Y_ε, Y_i internal to the device associated with the left- and right-hand terminal-pairs respectively. The total shunt admittance Y_{in}, measured at the input terminal with the output terminal shorted, is

$$Y_{in} = (i_{in}/\varepsilon)_{\varepsilon_{out} - 0} = Y_\varepsilon + Y.$$

The similarly defined output shunt admittance is $Y_{out} = Y_i + Y$.

The direct admittance, which is the defined admittance we are concerned with measuring, is $Y = (i/\varepsilon)_{\varepsilon_{out} = 0}$.

Y_ε and Y_i play no part in the definition of Y because ε is the potential difference found across the left-hand port, irrespective of how much current flows through Y_ε; and since $\varepsilon = 0$ across the right-hand terminal-pair, no part of the current i is shunted away through Y_i. This situation is in contrast to that of the previous section where Y_ε and Y_i would be directly added to, and become part of, the immittance of the device.

In §2.6 we discuss how conditions equivalent to this definition can be achieved and the effects of shunt admittances on four-arm bridge comparisons eliminated.

2.2.7 Three terminal definition

Whenever this phrase is encountered the intended meaning needs to be examined with especial care. The meaning customarily intended is a guarded or electrically screened two terminal device.

With reference to figure 2.8, the current i_ε which flows through the admittance Y_ε of the equivalent network from the component to the screen is directed elsewhere in the measurement network and plays no part in the definition of the impedance of the component. Devices defined in this way cannot be used in a current-equalised network. High-value resistors are often provided with three single terminals and their value is intended to be realised in accordance with the above definition. Some AC capacitors are defined this way and despite being provided with two coaxial fittings they cannot be used in a current-equalised network as the outer conductor of the high potential

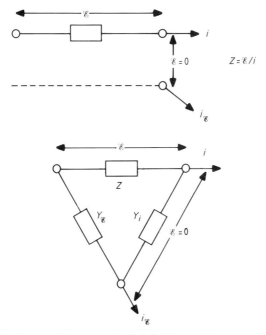

Figure 2.8 *The usual three terminal definition and its equivalent circuit.*

port is not connected to the outer of the lower and serves only as a termination of the screening of the high potential lead. The comparison of such devices is discussed in the older literature; from our point of view it would be better if they were reconstructed and defined in a logical manner as a degenerate two terminal-pair, as shown in figure 2.9.

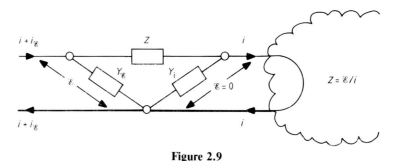

Figure 2.9

Usually, if two terminals are provided on the screening container of the component there is a low-impedance connection between them. If this impedance is negligible, the two terminals can be treated as if there were only one.

2.2.8 *Four terminal-pair definition*
In order to overcome both causes of uncertainty associated with the simple two terminal-pair definition it is necessary to combine the concepts of four terminal and two terminal-pair definitions. The result, shown in figure 2.10, is a four terminal-pair definition which is precise enough for all present practical applications and all values of components. It will be shown below that the uncertainty in the product of the impedances and admittances associated with the coaxial cables and connectors used to connect the devices into a network sets a limit to the precision.

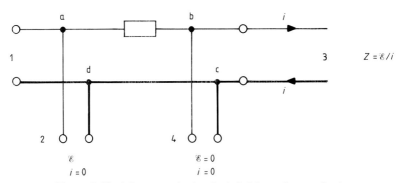

Figure 2.10 *A four terminal-pair definition of a standard.*

A simple way of appreciating the philosophy behind this definition is to recognise the assembly as two four-terminal devices. The current in the lower is equal and opposite to that in the upper. If there is no difference in potential between the internal defining points b and c, the potential between the points a and d together with the current i is clearly sufficient to define an impedance $Z = \varepsilon / i$ for the device. Remembering that no current should flow along potential leads, the measurement conditions to be fulfilled to realise the defined impedance Z of the device are given by the formal statements in the diagram.

The numbering of the terminal-pairs or ports follows that in the original paper of Cutkosky, and has become the standard notation for these devices.

A generator may be connected across the terminals of port 1 to provide the current through the device and another two terminal-pair or four terminal-pair device can be connected to port 3 for purposes of making a comparison. The defining conditions for the impedance of the device are not affected by the shunt impedances of other devices connected to ports 1 and 3—all that matters is that the inward and outward currents at the terminals of port 3 are identical and that the defining conditions at ports 2 and 4 are fulfilled. We anticipate later discussion by noting that, for example, port 2 could be connected to the known output potential of a transformer and a generator connected to port 1 adjusted until no current flows at port 2; the remaining defining conditions could be fulfilled by simultaneously bringing a detector connected to port 4 to a null.

2.3 Measuring four terminal-pair admittances in a two terminal-pair bridge by extrapolation

If the defining conditions at a port or terminal-pair are not fulfilled, we can nevertheless deduce the bridge balance condition corresponding to exact fulfilment by extrapolating the change produced on increasing the non-fulfilment. This is an example of extrapolation techniques which make use of the property of linearity.

With reference to figure 2.11, the admittance $Y = i / \varepsilon$ is fixed *completely* by the defining conditions at ports 2 and 4, provided that any cables between the standard and the remote ports 2 and 3 which give rise to Z_2, Y_2, Z_3 and Y_3 are considered to form part of the standard. We will call such cables 'defining cables'. Note that neither the internal impedance of the generator across port 1 nor the load impedance across port 3 matter if the defining conditions are satisfied.

If the defining conditions are not satisfied we can note the bridge balance both in this state, and again when known admittances have been added in turn to ports 1 and 4. Then an extrapolation to the balance which would have been obtained under the proper defining conditions can be made.

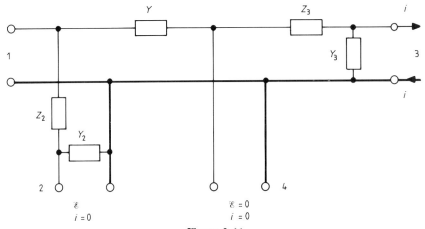

Figure 2.11

Consider the effect of an admittance added to port 1. In a two terminal-pair bridge, shown in figure 2.12, a generator is connected to port 2 (which has a defining cable) and a detector to port 3 (which again has a defining cable).

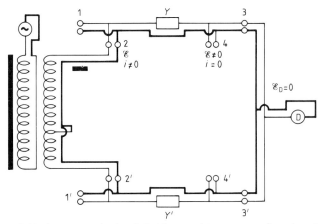

Figure 2.12 *A two terminal-pair bridge used to measure four terminal-pair admittances.*

As usual, the current returning via the outer conductors is made equal to that in the inner conductors by putting a current equaliser on the cable going to port 2 of Y.

Two admittances are added to the ports of each four terminal admittance

Y and Y'. At the high potential end an admittance, of value equal to that measured at port 2 with port 3 shorted, is connected across port 1. It will be seen that this doubles the current taken from the generator at port 2, whilst leaving the four terminal-pair admittance unaltered. Extrapolation is therefore equivalent to achieving the defining conditions at port 2.

A second admittance, equal in value to that measured at port 3 with port 2 shorted, is connected across port 4. Its action is more subtle, and needs some analysis to explain. Note that ports 2 and 2' can both be 'shorted' by removing the supply and shorting the primary of the transformer. The measured value of the admittance looking into 3 with 2 shorted is, to a sufficiently good approximation, $Y + Y_3$. To see the action of an admittance of this value added to port 4, we examine the following cases: (i) where the four terminal-pair conditions are strictly fulfilled, (ii) the actual measurement conditions, i.e. without the extra admittance added to port 4 and (iii) with this extra admittance added.

(i) With reference to figure 2.13 where $\varepsilon = 0$, $i = 0$ at port 4, the current at the internal defining junction b is $i = \varepsilon Y$. At port 3 however, because of the shunt Y_3, across which a potential iZ_3 exists, the current is

$$i + i' = \varepsilon Y + \varepsilon Y Z_3 Y_3 = \varepsilon Y(1 + Z_3 Y_3). \tag{2.1}$$

(ii) With zero potential across port 3, as in figure 2.14, the current at port 3, through Z_3 and Y in series, is

$$i = \varepsilon (1/Y + Z_3)^{-1} = \varepsilon Y/(1 + YZ_3). \tag{2.2}$$

The amount by which the current at port 3 is in error is given by the difference of equations (2.2) and (2.1):

$$\delta i = -\varepsilon Y Z_3 (Y + Y_3) \tag{2.3}$$

neglecting second-order terms.

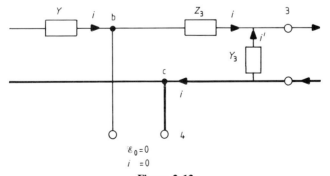

Figure 2.13

(iii) With zero potential across port 3 and a shunt of $Y + Y_3$ across port 4, as in figure 2.15, part of the current i' which flows from the source through Y is diverted through $Y + Y_3$ at port 4, leaving only i to flow at port 3. No current flows through Y_3 since $\mathcal{E}_d = 0$ there.

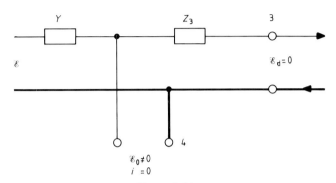

Figure 2.14

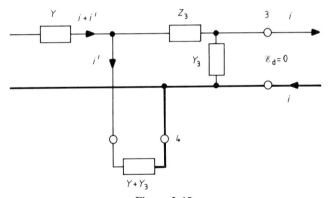

Figure 2.15

We have

$$\mathcal{E} = (i + i')/Y + iZ_3$$

and

$$iZ_3 = i'/(Y + Y_3)$$

whence, eliminating i',

$$i = \mathcal{E}\,Y/(1 + Z_3 Y_3 + 2Z_3 Y). \qquad (2.4)$$

Again, the amount by which this current is in error is given by the difference of (2.4) and (2.1) as

$$\delta i = -2\varepsilon YZ_3(Y + Y_3). \qquad (2.5)$$

We note that equation (2.5) is just twice equation (2.3), so that adding the shunt $Y + Y_3$ has doubled the error current, and hence extrapolation yields the condition for zero error current, or the correct four terminal-pair defining condition.

It is not necessary to use the particular values of shunt admittances suggested, provided the values are known. Let the ratios of the shunt admittances added to ports 1 and 4 to those measured at ports 2 and 3 with ports 3 or 2 respectively shorted be k_1 and k_2; then the changes of bridge balances brought about by adding such shunt admittances in turn to the appropriate ports can be weighted by $k_{1,2}$ and added algebraically to give the total correction to the bridge balance.

2.4 The effect of cables connected to the ports of impedance standards

So far we have implicitly assumed that the conditions obtained at the internal defining points of a device hold also at the ports even though there may be a considerable length of conductor between them; there is also the question of the finite length of conductor required to connect the devices into a practical bridge network. In this book we are predominantly concerned with coaxial conductors used to make such connections, and so we will not consider any other type of connection, as, for example, twisted wire-pairs.

A coaxial cable will have a shunt admittance because of its capacitance between the inner and outer conductors and these conductors will possess finite series impedances because of their resistances and self-inductances. Remembering that we are considering only the approximation where the lengths of cable are negligibly short compared with the wavelength of the measurement frequency, this admittance and these impedances can be considered as 'lumped', that is, as if the measured immittances of the cable acted at a single point along an otherwise immittance-free cable. Further, we can analyse the effect of cables as if the impedance of the outer conductor is zero and only that of the inner is appreciable. This is because, in the circuits we describe, the currents in inner and outer conductors are constrained to be equal and opposite so that the change of the voltage difference between inner and outer between two points along a cable can be ascribed to an impedance which is the algebraic sum of the impedance of each conductor individually in the presence of the other. Therefore, provided that we remember that the impedance of the inner discussed in this section is to be in reality the sum of the impedances of inner and outer, the results derived here can be applied directly.

By the admittance of a cable we mean the capacitance and shunt conductance measured at either end of the cable with the other end open-circuited; strictly, to eliminate end effects a difference measurement should be made with the cable inserted between the measuring instrument and a dummy lead of identical cable, but such effects are usually negligible to the accuracy with which we need to know Y.

Similarly, by the impedance Z we mean the resistance and series inductance measured at one end with the other end short-circuited; again, end effects could be eliminated if required by inserting the cable (unshorted) between the measuring instrument and a dummy shorted length which is again measured separately and the difference taken.

We take the circuit shown in figure 2.16 as the equivalent circuit of a cable. We shall find that, even if we regard the impedance of a standard as having been redefined with reference to new exit ports at the ends of the added cables, uncertainties still arise because of variable connector immittances at these and the original ports. The successive elaboration of the above definitions of the impedance of a standard is largely an attempt to minimise this effect.

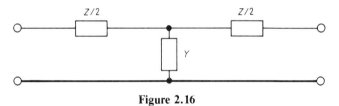

Figure 2.16

Two special cases of conditions obtained at one end of a cable under specified conditions at the other are of interest to us.

(i) The relation between the current at one end of a cable and the current at the other under conditions of zero potential there. Note, in figure 2.17, that the network in the right-hand 'cloud' must contain sources and be capable of setting up the specified conditions; the left-hand termination could be a short-circuit, but is not necessarily so.

The sum of the currents at the network junction a is zero, hence the current through Y is $i' - i$. Traversing the left-hand mesh,

$$i'Z/2 + (i' - i)/Y = 0$$

that is:

$$i = i'(1 + YZ/2).   (2.6)$$

(ii) The dual situation, which is the relation between the potential at one end of the cable and that at the other under conditions of zero current there. Again, as in figure 2.18, the networks in the clouds set up the

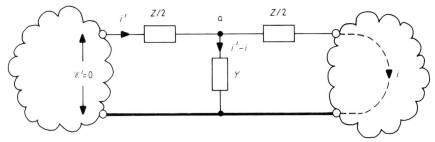

Figure 2.17

required conditions; the termination on the right could be an open circuit, but is not necessarily so. Traversing the left-hand mesh,

$$\mathcal{E} = iZ/2 + i/Y = i(Z/2 + 1/Y)$$

and the right-hand mesh, where $\mathcal{E}' = i/Y$, hence

$$\mathcal{E} = \mathcal{E}'(1 + YZ/2). \tag{2.7}$$

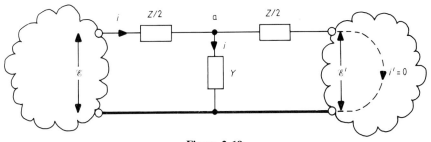

Figure 2.18

In contrast with the original paper of Cutkosky, our representation of a cable has the impedance Z divided in equal halves at the internal point a. This gives rise to the factor of two in our expression where Cutkosky has unity. The reason for our representation can be seen if we consider the above cases when the clouds are connected by two similar cables of admittances $k_1 Y$, $k_2 Y$ and impedances $k_1 Z$, $k_2 Z$ respectively (as in figure 2.19):

$$i' k_1 Z/2 + (i' - i_1)(k_1 Y)^{-1} = 0$$

$$-(i' - i_1)(k_1 Y)^{-1} + i_1(k_1 + k_2)Z/2 + (i_1 - i)(k_2 Y)^{-1} = 0.$$

Since $YZ \lesssim 10^{-6}$ for moderate lengths of the order of a metre of cable at the low frequencies with which we are concerned, we may neglect the second-order terms, leaving

$$i = i'[1 + (k_1 + k_2) Y(k_1 + k_2)Z/2 + O(YZ)^2].$$

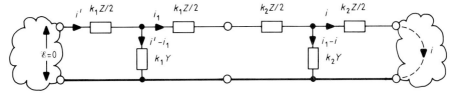

Figure 2.19

Comparing the result with equation (2.6) above we see that the combined cable behaves as if it had total lumped impedances $(k_1 + k_2)Z$ and admittances $(k_1 + k_2)Y$ and this is what, to a sufficient approximation, would be obtained by a direct measurement. This confirms the validity of our equivalent T-network for a cable.

2.4.1 The effect of cables on a two terminal component
With reference to figure 2.20,

$$\mathcal{E}' = i'Z_c/2 + (i' - i)/Y_c \tag{2.8}$$

$$\mathcal{E} = (i' - i)/Y_c - iZ_c/2 \tag{2.9}$$

but $\mathcal{E}/i = Z$, so equation (2.9) becomes

$$i/i' = (1 + Z_c Y_c/2 + ZY_c)^{-1}$$

whence, from equation (2.8)

$$\mathcal{E}'/i' = Z_c/2 + (Z_c/2 + Z)/(1 + Z_c Y_c/2 + ZY_c).$$

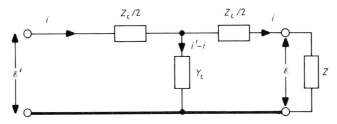

Figure 2.20 *A cable added to a two terminal component.*

Clearly, even if the YZ terms are negligible, there is a first-order addition of Z_c to Z. If we view a connector as a 'cable' between Z and a network, even substitution measurement will be in error by the uncertain connector impedance Z_c and if Z is a large impedance the term $Y_c Z$ may be appreciable for even a short length of cable or a good quality connector. Two terminal standards with a coaxial port are only useful when moderate accuracy is required over a limited range of values.

2.4.2 The effect of cables on a two terminal-pair connection of a four terminal component

A good definition of the arrangement of §2.2.5 is obtained when connecting a component of this kind to a network with coaxial cables only if one cable carries the current to and from the component and the potential difference is maintained between inner and outer conductors of the other cable.

Drawing the standard with attached cables as in figure 2.21, from the left-hand mesh

$$(i+i_1-i')/Y_1+(i+i_1)Z_1/2+iZ=0 \qquad (2.10)$$

and from the right-hand mesh,

$$i_1(Z_2/2+Y_2^{-1})-iZ=0. \qquad (2.11)$$

Eliminating i_1 between equations (2.10) and (2.11),

$$i'=i[1+(ZY_2+ZZ_1Y_1Y_2)(1+Y_2Z_2/2)^{-1}+Y_1Z_1/2+Y_1Z]$$

from equation (2.7) above,

$$\mathcal{E}'=\mathcal{E}(1+Y_2Z_2/2)^{-1}.$$

Hence the four terminal admittance sensed at the ends of coaxial cables is

$$1/Z'=i'/\mathcal{E}'=Z^{-1}(1+Y_1Z_1/2+Y_2Z_2/2+Y_1Z+$$

$$Y_2Z+Y_1Y_2ZZ_2/2+Y_1Y_2ZZ_1/2+Y_1Y_2Z_1Z_2/4)$$

since $i/\mathcal{E}=Z^{-1}$. The product correction terms can readily be evaluated from the known properties of the cables and the impedance Z to be measured.

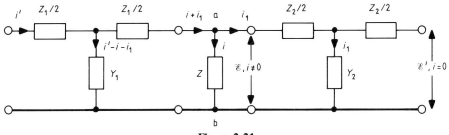

Figure 2.21

Typically, for a metre of coaxial cable of 50 Ω characteristic impedance

$$Z_2=10^{-2}+j\omega2.5\times10^{-7} \qquad Y_2=+j\omega10^{-10}$$

($L_2=0.25\,\mu\text{H m}^{-1}$, $C_2=100$ pF m^{-1}, $R_2=0.01\ \Omega\,\text{m}^{-1}$) and the term $Y_2Z_2/2$ is then $-\omega^2 2.5\times10^{-17}+j\omega10^{-12}$.

At frequencies below $\omega = 10^5$ rad s^{-1} ($\simeq 16$ kHz) this term will not exceed 3 parts in 10^7, and neither will the term $Y_1 Z_1/2$ if the potential cable is similar. The term $Y_1 Y_2 Z_1 Z_2$ is usually quite negligible. These terms are independent of the value of the impedance Z being measured, but the remaining second-order term $Y_1 Z$ and fourth-order terms of course are not.

We consider three special cases, taking the cable to have the properties listed above.

(i) Z is a pure capacitance, C.

$$Y_1 Z = \mathrm{j}\omega 10^{-10}/\mathrm{j}\omega C = 10^{-10}/C$$

as expected, because C is merely shunted by the cable capacitance. Clearly, the terminal configuration considered in this section is unsuitable for all but very large values of capacitance.

(ii) Z is a pure inductance, L.

$$Y_1 Z = \mathrm{j}\omega 10^{-10} \times \mathrm{j}\omega L = -\omega^2 10^{-10} L.$$

The use of this terminal configuration is not ruled out for moderate values of ω and L. For example, at $\omega = 10^4$ rad s^{-1} ($\simeq 1.6$ kHz) and $L = 10^{-3}$ H, $Y_1 Z = -10^{-5}$. This is an almost negligible correction to this type of standard whose accuracy is usually limited by both a lack of stability and a large external magnetic field.

The term $Y_1 Y_2 Z Z_2$ is $-\mathrm{j}\omega^3 10^{-20}(10^{-2} + \mathrm{j}\omega 2.5 \times 10^{-7})L$, and again for moderate values of ω this is smaller than the term just evaluated.

(iii) Z is a pure resistance R.

$$Y_1 Z = \mathrm{j}\omega 10^{-10} R$$

and this is in quadrature with R, but at $\omega = 10^4$ rad s^{-1} ($\simeq 1.6$ kHz) it becomes appreciable for values of R which exceed 10^3 Ω if the cable shunt admittance constituting Y_1 is an only moderately lossy capacitance. If this capacitance has a phase angle of 10^{-3} this term will give an in-phase addition to R of $\omega 10^{-10} \times 10^{-3} R$ which again at $\omega = 10^4$ rad s^{-1} is $10^{-9} R$. Therefore this terminal arrangement is a poor way of defining R for accurate measurements at this or similar frequencies if the value of R is greater than 1 kΩ.

The fourth-order term $-Y_1 Y_2 Z Z_2$ is again much smaller for moderate values of ω than the term just evaluated.

2.4.3 The effect of cables on a two terminal-pair component

This case is shown in figure 2.22 where the defining conditions are again to be fulfilled at the ends of added cables instead of at the internal defining points. From equation (2.6) above, we have $i = i'(1 + Y_2 Z_2/2)$. Since current flows through the original terminals where the potential was defined we cannot use equation (2.7) to deduce the relationship between ε and ε'. We can carry out an approximate analysis, however. If Y_2, Y_1 are less than, or of the order of, Z^{-1} then the current through Y_ε is $\varepsilon Y_\varepsilon$ and the current through Y_1 is approximately εY_1. Therefore

$$\varepsilon' - \varepsilon = Z_1(\varepsilon Y_1 + \varepsilon Y_\varepsilon + i)/2 + Z_1(\varepsilon Y_\varepsilon + i)/2$$

or

$$\varepsilon' = \varepsilon(1 + Y_1 Z_1/2 + Y_\varepsilon Z_1 + Z_1/Z).$$

Since $Z = \varepsilon/i$ and the apparent impedance of the component and added cables $Z' = \varepsilon'/i'$,

$$Z' = Z(1 + Y_1 Z_1/2 + Y_\varepsilon Z_1 + Z_1/Z)(1 + Y_2 Z_2/2).$$

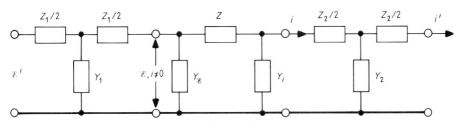

Figure 2.22

Hence, the correction to be made to the apparent impedance Z' to obtain Z depends to first order on the term Z_1/Z, which might have been anticipated since Z_1 is in series with Z. It is subject to the uncertainty in connector impedance contained in Z_1. Either satisfactory corrections can be made for account for the other two terms, or the impedance can be regarded as having been redefined with the lengths of additional cable included. Because of these considerations, a two terminal-pair definition is suitable only for high impedances Z.

Since the way we define a three terminal impedance is not different in principle from a two terminal-pair definition, the above considerations apply.

2.4.4 The effect of cables on a four terminal-pair component
The addition of a cable between the generator network and port 1 has no significance as no quantities are defined at this port. A cable added to port 4 also has no effect since if $\varepsilon = 0$ and $i = 0$ at one point on a cable, these conditions must hold everywhere on it and therefore its length is immaterial. We need only consider the effect of adding cables to ports 2 and 3, as shown in figure 2.23.

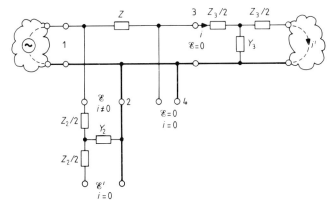

Figure 2.23

Using equation (2.6) we have (reversing the roles of i and i')

$$i' = i(1 + Y_3 Z_3/2)$$

and from equation (2.7)

$$\varepsilon' = \varepsilon(1 + Y_2 Z_2/2)^{-1}.$$

Hence the apparent impedance Z' after the addition of the cables is simply

$$Z' = \varepsilon'/i' = Z(1 + Y_2 Z_2/2)^{-1}(1 + Y_3 Z_3/2)^{-1}.$$

Notice that the correction terms are now much simpler. It is especially important to note that they do not involve the impedance Z. This means that a four terminal-pair definition is suitable for *all* values of Z, that is, for all components at all frequencies within our restriction that the wavelength is negligible compared with cable lengths. The effects of variable connector impedance, which we can view as being contained in Z_2, Z_3, although still present, are the same for all Z. By making a sensible choice of connector types and restricting the values of Y_2, Y_3, the uncertainty in the value of Z can be kept less than a part in 10^9 at frequencies below 16 kHz.

For metre lengths of typical cable at 1.6 kHz the correction terms are of the order of 1 in 10^9; they are easily calculated from the measured properties of the cables, but more usually Z is taken as having been redefined at the new ports at the ends of the added cables which become part of the total defined impedance Z'. This of course implies that standard cables of a fixed length are used for the purpose: a change in length of 100 mm in cables having the constant impedance and admittance per unit length assumed above changes the four terminal-pair admittance by 1 in 10^{10}.

Although we shall find the complexity of bridges to compare four terminal-pair impedances is increased, the use of such impedances is to be encouraged, and in work of the highest accuracy it is essential.

2.5 The relationship between a conductor-pair network and the more familiar single conductor network_____

In this book we are concerned with conductor-pair networks where one network of the pair has a low impedance compared with the other, and where, by using one current equaliser in each mesh of the network, the current in any one conductor is matched by an equal and opposite current in the other conductor of the pair. The potential differences between corresponding nodes of the pair of networks are quantities of interest because, for example, detectors or sources are connected between corresponding nodes.

A conductor-pair network can easily be transformed to the more familiar corresponding single conductor network in order to appreciate the relationship between the two situations. Consider a portion of the equalised conductor-pair network shown in figure 2.24.

$$(\mathcal{E}_A - \mathcal{E}_{A'}) - (\mathcal{E}_B - \mathcal{E}_{B'}) = (\mathcal{E}_A - \mathcal{E}_B) - (\mathcal{E}_{A'} - \mathcal{E}_{B'}) = i(Z+z).$$

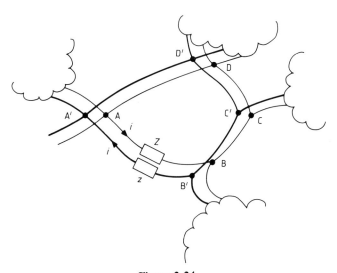

Figure 2.24

If we replace the primed net with one of zero impedance ($\mathcal{E}_{A'} = \mathcal{E}_{B'}$) so that the entire new primed net is an equipotential, and enhance every impedance between two nodes of the unprimed net by the corresponding impedances of the old primed net, a single net with the same nodal equations $\mathcal{E}_A - \mathcal{E}_B = i(Z+z)$ as the conductor-pair net is obtained, and is therefore equivalent to it. The potentials $\mathcal{E}_A$, $\mathcal{E}_B$ etc will be referenced to some arbitrary zero, such as the new primed net.

In general, as we have already noted, each z has a much smaller value than the corresponding Z.

2.6 An analysis of conductor-pair bridges to show how the effect of shunt admittances can be eliminated_____

In this section we examine how a bridge network can compare the direct admittances of devices whilst ignoring their input and output shunt admittances. (These quantities have been defined in §2.2.6.) It is possible to so arrange components that the bridge balance condition can be made to depend mostly on the *direct* admittances but with a correction term which involves the shunt admittances. This term can be factorised into the product of two other terms, both of which depend on shunt admittances, but both of which can be made small by subsidiary adjustments. The product is then usually negligible. We will discuss two terminal-pair components for simplicity, as this aspect of bridge technique is not significantly different for devices defined in a four terminal-pair manner; we will take just the three most commonly occurring cases, namely the voltage transformer ratio bridge, the classical general four-arm bridge and the quadrature bridge which is used for the comparison of resistors and capacitors. Some elements of the discussion will be recognised as the Wagner and detector auxiliary balances of the older bridge techniques applied to better defined conductor-pair measurements.

Thompson (1964) has also given an elegant general treatment of this question.

2.6.1 Comparing direct admittances with voltage sources
Consider the arrangement shown in figure 2.25 where two terminal-pair devices are connected to separate sources. Their direct admittances Y_1 and Y_2

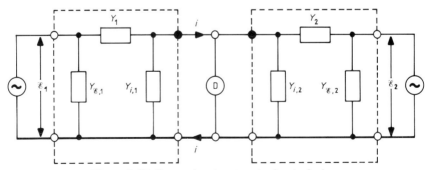

Figure 2.25 *Comparing two terminal-pair devices.*

are to be compared by adjusting ε_1, ε_2 or the ratio $\varepsilon_1/\varepsilon_2$ to null the detector.

If the sources are of negligible output impedance (for example, if they are the outputs of two separate windings of a well-constructed voltage ratio transformer) their output potential differences ε_1 and ε_2 are not affected by the presences of the shunt admittances $Y_{\varepsilon,1}$ and $Y_{\varepsilon,2}$ which are just loads on the sources. Therefore, to a good approximation $Y_{\varepsilon,1}$ and $Y_{\varepsilon,2}$ do not affect the bridge balance. Also, when the detector is nulled, Y_{i1}, Y_{i2} have no potential across them. Hence no current flows through these admittances, and their value is also immaterial to the bridge balance condition which depends only on the well-defined direct admittances Y_1 and Y_2. Because $Y_{i1,2}$ shunt the detector terminals, however, they do degrade the apparent detector sensitivity and noise performance. (We assume in this section that the series impedances of the connections from the two terminal-pair devices to the detector are negligible so that $\varepsilon = 0$ at the terminals of the devices as well as at the detector terminals. The form of the error term involved in this approximation has already been discussed in §2.4.1 in connection with the shunt admittance of an attached cable.)

2.6.2 Comparisons made with a four-arm bridge

The situation is only slightly more complicated if the ratio of voltages $\varepsilon_1 : \varepsilon_2$ is not established with a voltage ratio transformer but by the ratio of two other two terminal-pair admittances. The network, which is drawn in figure 2.26, is a classical four-arm bridge needing the addition of Wagner and detector auxiliary networks in order to make accurate measurements.

The ratio of voltages $\varepsilon_1 : \varepsilon_2$ is established by the variable Wagner or source balance components $Y_{c,d}$ acting in conjunction with the rest of the bridge network. If we view the network as a combination of two networks, as in the preceding section, one of the two detectors, D_a, can be retained but D_b can be replaced by D_m nulled simultaneously with D_a. D_m is the familiar main detector of the classical four-arm bridge. Reciprocity (the unchanged behaviour of a bridge on interchanging source and detector) then suggests that D_m should also be provided with adjustable shunt admittances $Y_{a,b}$ as drawn. The values of $Y_{a,b,c,d}$ include the contiguous shunt admittances of the main bridge components. We seek to show that by suitable adjustment of the auxiliary components $Y_{a,b,c,d}$ the problem of the finite internal impedance of the 'generators' $\varepsilon_{1,2}$ can be overcome.

Suppose that by a suitable choice of values for the adjustable ratios $Y_3 : Y_4$ (main balance), $Y_a : Y_b$ (detector balance) and $Y_c : Y_d$ (source balance) both detectors $D_{a,m}$ are nearly nulled. If then the source balance is altered grossly from this condition by, for example, shorting Y_c, a nulled condition of D_m can be achieved by adjusting the $Y_a : Y_b$ ratio so that the nodes a and b have the same potential. The potential differences between these nodes and the corresponding nodes a' and b' of the outer circuit will then be the same, and

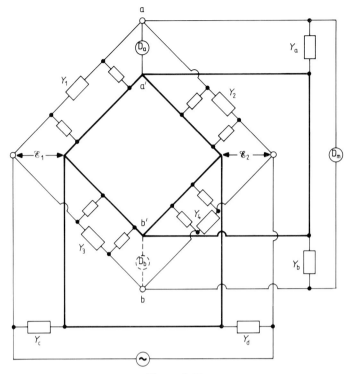

Figure 2.26

the criticality of adjustment of the $Y_a:Y_b$ ratio will be proportional to the magnitude of this potential difference.

It will be seen that Y_a in parallel with Y_1, Y_b in parallel with Y_3, Y_2 and Y_4 form a bridge with respect to $\mathcal{E}_2$ as source and D_m as detector whose balance condition is

$$(Y_1 + Y_a)/(Y_3 + Y_b) = Y_2/Y_4. \tag{2.12}$$

The potential difference between a and a', however, can be made very small by, say, shorting Y_b and adjusting D_a to a null by altering the source ratio Y_c/Y_d because Y_c in parallel with Y_3, Y_d in parallel with Y_4, Y_1 and Y_2 from a bridge, having D_a as a detector, with respect to the actual source. Its balance condition is

$$(Y_3 + Y_c)/(Y_4 + Y_d) = Y_1/Y_2. \tag{2.13}$$

It is not necessary to use a second detector D_a. Instead Y_b can be temporarily shorted, making D_m equivalent to D_a.

Finally, removing the short across Y_c produces a condition where the reading of D_m is proportional to the product of the accuracy of source and detector balances which include all the effects of the shunt admittances. One in 10^4 accuracy in each of these balances is sufficient to achieve 1 in 10^8 accuracy of comparison of the *direct* admittances appearing in the main balance condition

$$Y_1/Y_2 = Y_3/Y_4. \tag{2.14}$$

Algebraic manipulation of equations (2.12) and (2.13) with equation (2.14) yields the simpler auxiliary balance conditions

$$Y_a/Y_b = Y_2/Y_4 \tag{2.12'}$$

and

$$Y_c/Y_d = Y_1/Y_2. \tag{2.13'}$$

In other words, we have achieved a condition where even if the source ratio $\varepsilon_1:\varepsilon_2$ is not correct, the detector does not respond to it; and conversely, if the detector does respond, the source balance is adjusted to get the ratio $\varepsilon_1:\varepsilon_2$ correct. Therefore one obtains the benefit of the product of the two adjustments in immunising the main balance condition against them, and the effects of the shunt admittances of the components $Y_{1,2,3,4}$ are eliminated. Also, neither the components $Y_{a,b,c,d}$ nor the shunt admittances which partially comprise them need to be nearly as stable as the direct admittances $Y_{1,2,3,4}$ being compared.

2.6.3 Comparisons made with a quadrature bridge
The four-arm bridge deals with the situation

$$Y_1 Y_4 / Y_2 Y_3 = +1$$

but, particularly for comparing the values of resistors and capacitors, the condition

$$Y_1 Y_4 / Y_2 Y_3 = -1$$

has advantages.

This condition can be realised in the network shown in figure 2.27 where, for simplicity, we have omitted showing explicit the shunt admittances of the main components $Y_{1,2,3,4}$ and the complete circuit of the outer conductors. This network is known as a quadrature bridge and is used to relate the impedance of a resistance to that of a capacitance at a known frequency. This operation is an essential step in relating the practical resistance unit, the ohm, to the unit of capacitance, the farad, as realised in SI units by a calculable capacitor.

D_1 is nulled if $\varepsilon_1 Y_1 = \varepsilon_2 Y_2$. D_2 is nulled if $-\varepsilon_1 Y_4 = \varepsilon_2 Y_3$. Hence *one* condition for both being nulled is

$$-Y_1/Y_4 = Y_2/Y_3. \tag{2.15}$$

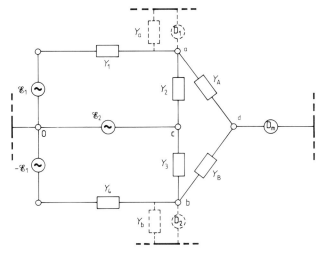

Figure 2.27

We can combine D_1 and D_2 using Y_A and Y_B to go to a single detector D_m in such a way that the *other* condition, the actual value of $\mathcal{E}_2/\mathcal{E}_1$, that is, the magnitude of $\mathcal{E}_2$, does not have to be strictly correct.

The superposition theorem enables us to consider the effect on D_m of $\mathcal{E}_2$ alone, as if $\pm\mathcal{E}_1$ were zero. $\mathcal{E}_2$ will cause a potential difference between bridge nodes a and b, but D_m will not respond to this, if, considering the bridge Oadb,

$$Y_A/Y_B = (Y_1 + Y_a)/(Y_4 + Y_b).\qquad(2.16)$$

If, further, Y_A and Y_B are adjusted to achieve this condition, nodes a and b will be at the same potential, if, considering the bridge aObc,

$$Y_2/Y_3 = (Y_1 + Y_a)/(Y_4 + Y_b).\qquad(2.17)$$

Condition (2.16) is termed the detector balance and (2.17) is termed the j balance, because in the practical R–C bridge, $\mathcal{E}_2$ is in quadrature with $\mathcal{E}_1$.

Hence, if neither condition is fulfilled exactly, the residual current through the detector will be proportional to the product of the two imbalances.

Again, neither $Y_{a,b,A,B}$ need be as stable as the main bridge admittances $Y_{1,2,3,4}$.

A quadrature bridge has two capacitors and two resistors as its main components, connected as in figure 2.28. To null the two detectors, with $1/R_2 = 1/R_4 = \omega C_1 = \omega C_3$ we need $\mathcal{E}_2 = -\text{j}\mathcal{E}$. In accordance with the above analysis, we need a way of combining $\mathcal{E}_a$ and $\mathcal{E}_b$ so that one detector responds to either of them, but not to small variations in $-\text{j}\mathcal{E}$. This can be accomplished with the components shown in figure 2.29, where ωC_a, ωC_B,

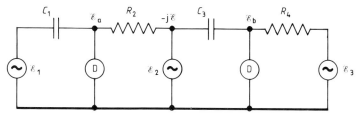

Figure 2.28

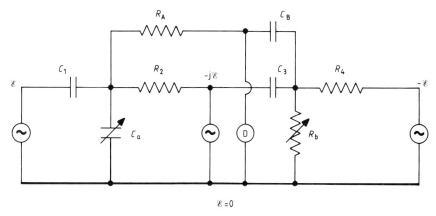

$\mathcal{E} = 0$

Figure 2.29

$1/R_A$, $1/R_b$ and the main components ωC_1, ωC_3, $1/R_2$, $1/R_4$ are all nominally equal in magnitude.

The superposition theorem enables the response of the detector to $-j\mathcal{E}$ to be calculated, by setting $\mathcal{E}$ and $-\mathcal{E}$ to zero and calculating the response to $-j\mathcal{E}$ alone.

With respect to $-j\mathcal{E}$ and the detector, the network can then be recognised as a 'twin-tee' bridge which can be balanced by altering C_a and R_b (which may in practice need some shunt conductance and capacitance respectively to account for small quadrature defects in the bridge components) so that the detector does not respond to $-j\mathcal{E}$ at all. This is the adjustment for the detector network balance. Then C_a, say, can be shorted, $\mathcal{E}$ and $-\mathcal{E}$ restored, and $-j\mathcal{E}$ adjusted for a detector null. This is the source balance. Shunting the main components with adjustable lower value admittances or putting adjustable small voltage sources in series with $\pm\mathcal{E}$ permits the main balance to be made.

Unfortunately, the presence of R_A in the detector network increases the Johnson noise of the network as seen by the detector somewhat, but the reduction of the dependence of the bridge balance to only the product of the subsidiary balances is an overriding requirement in practice.

Nominal values used in some practical quadrature bridges are as shown in figure 2.30. The voltage $-j\mathcal{E}$ is obtained from the outer $10^5\ \Omega$ resistor and 1 nF capacitor. If there were an open circuit at their junction the voltage there would be $-2j\mathcal{E}$, but the admittance of the remainder of the network at the node where $-j\mathcal{E}$ is required is a parallel combination of a $10^5\ \Omega$ resistor and a 1 nF capacitor; the loading effect of these halves the open circuit voltage $-2j\mathcal{E}$ whilst leaving the phase unaltered.

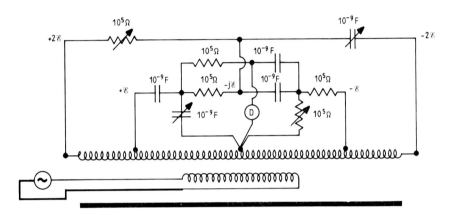

Figure 2.30 $\omega = 10^4\ \text{rad s}^{-1}$ *(the admittance of 10^{-9} F equals that of $10^5\ \Omega$ at $\omega = 10^4\ \text{rad s}^{-1}$).*

The iterative balance procedure is to adjust the detector to a null by

(i) adjusting the source balance with the detector balance completely offset by, for example, temporarily shorting one of its adjustable components,

(ii) similarly temporarily shorting one of the components of the source network and adjusting the detector coupling network,

(iii) adjusting the main balance.

The sequence (i), (ii) and (iii) is then repeated until no further adjustments are needed for all three balance conditions to be successively obtained to the required accuracy. (The adequacy of the (i) and (ii) operations can readily be checked at any time by observing (iii) whilst making a change which is a little larger than their observed uncertainty of adjustment in (i) and (ii) successively.) This procedure will be recognised as being of the utmost utility in any bridge which requires balancing of multiple detectors.

The fact that the bridge is based on a $1:-1$ voltage ratio can be used to eliminate any small inequality in the $\pm\mathcal{E}$ sources by simply interchanging the connections to them and taking the mean of the results obtained in the original and reversed configurations.

If the main components have small quadrature components, G_1, ωC_2, G_3, $\omega C_4 \ll \omega C_1$, $1/R_2$, ωC_3, $1/R_4$ respectively, such that

$$Y_p = j\omega C_p + G_p \qquad p = 1, 2, 3 \text{ or } 4$$

and G_2, $G_4 = 1/R_2$, $1/R_4$ respectively, the balance condition $Y_1 Y_3 / Y_2 Y_4 = -1$ becomes

$$\omega^2 C_1 R_2 C_3 R_4 [1 - G_1 G_3 / \omega^2 C_1 C_3 - j(G_1/\omega C_1 + G_3/\omega C_3)] =$$
$$1 - \omega^2 C_2 R_2 C_4 R_4 + j(\omega C_2 R_2 + \omega C_4 R_4).$$

If the products of small quantities $G_1 G_3 / \omega C_1 \omega C_3$ and $\omega C_2 R_2 \omega C_4 R_4$ are small enough to be neglected, the in-phase part of the above relation yields the simple balance condition

$$\omega C_1 R_2 \times \omega C_3 R_4 = 1 \qquad (2.18)$$

and the quadrature part the relation between the phase defects of the components,

$$G_1/\omega C_1 + R_2 \omega C_2 + G_3/\omega C_3 + R_4 \omega C_4 = 0. \qquad (2.19)$$

2.7 Combining networks to eliminate the effect of unwanted potential differences

2.7.1 General statement of the problem

If we require a linear combination of potentials $\mathcal{E}_a$ and $\mathcal{E}_b$ at nodes a and b respectively to appear at a third node c of a bridge network we can connect adjustable impedances Z_a and Z_b from a and b respectively to c. This additional mesh will of course modify the original potentials at a and b to, say, $\mathcal{E}'_a$, $\mathcal{E}'_b$. Nevertheless, values of Z_a, Z_b can be found which produce the required potential at c, although the necessary values depend on the whole of the network. Examples of this principle have already been met in §2.6. One particular instance in which this concept is applied and which appears frequently in immittance comparison bridges is described in this section.

With reference to figure 2.31, we suppose that the presence, or departure from a desired value, of some auxiliary source or sources of EMF $\mathcal{E}$ or potential iZ in a network, causes a potential difference between two nodes a and b which we wish to be at the same potential with respect to another node d. Since, for a linear network, the superposition theorem states that the effects of all sources can be considered separately and independently, we can consider the situation as equivalent to that in which the EMFs of all the other sources are set to zero. Then the two nodes a and b can each be connected by impedances Z_a and Z_b to a third node c and the values of these impedances adjusted until the potential of c with respect to d is *independent of the magnitude of the auxiliary sources*. That is, the presence of these sources

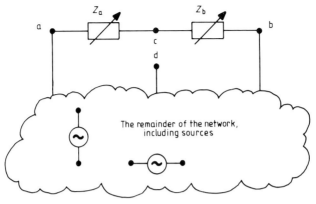

Figure 2.31

does not affect the potential of c with respect to d. c is therefore a node equivalent in potential to that a and b would have if the potential difference between them were zero.

It is evident that the added impedances and the original network form a balanced bridge, the potential difference of whose 'detector' nodes c and d does not respond to the 'source', the potential difference between a and b. The added impedances and the node c are often termed a 'combining network', and the concept is of the greatest value in designing bridge networks.

2.7.2 Illustration of the concept

In designing bridges containing four terminal or four terminal-pair impedances we need to connect two such components in series and yet have a condition equivalent to zero potential at the low-potential terminals or between the low-potential terminal-pairs of each component simultaneously. This, without the resource described above, is clearly not possible as the potential drop along the leads connecting the 'low' current terminals of the devices will cause the 'low' potential terminals to be at different potentials.

The simplest situation in which a combining network is used occurs in the DC network known as a 'Kelvin double bridge' illustrated in figure 2.32. This network is used to find the ratio of two low-value four terminal resistors R_1 and R_2 in terms of the ratio of two comparatively high-value two terminal resistors R_3 and R_4. We may note in passing that the potential terminals g and h of R_1 and R_2 have to supply the small but finite current which traverses R_3 and R_4, so that the defining conditions of R_1 and R_2 are only approximately fulfilled, but we are not concerned here with this imperfection.

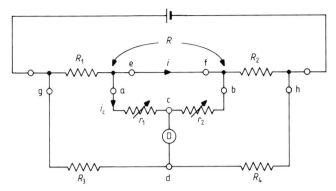

Figure 2.32 *Kelvin's double bridge. An example of the use of a combining network.*

The current i traversing R_1 and R_2 causes a potential difference iR (where R is the resistance between their 'low' internal defining points) between the 'low' potential terminals a and b. r_1 and r_2 form a combining network whose junction c is connected to the bridge detector. Values of r_1 and r_2 can be found such that the detector does not respond to changes in the potential difference between a and b brought about, for example, by temporarily opening the link between e and f. Once this is achieved, the actual situation is the same as that in which the potential difference between a and b is truly zero, equivalent to the defining condition $\varepsilon = 0$, $i = 0$ at the 'low' potential terminals of R_1 and R_2.

r_1 and r_2 together with the rest of the network form a bridge, balanced so that the detector does not respond to the 'source' iR. If the main balance condition $R_1/R_2 = R_3/R_4$ has also been achieved, their values are such that

$$r_1/r_2 = R_1/R_2 = R_3/R_4. \qquad (2.20)$$

That is, if the voltage developed across the resistance R is divided by the resistors r_1 and r_2 at the point c in the same ratio as the voltages across R_1 and R_2, the bridge balance is independent of the value of R, and is therefore identical with that which would be attained if the adjacent internal defining points of R_1 and R_2 were actually at the same potential. Warshawsky (1955) has discussed a network which uses four combining networks to allow full four terminal definitions of $R_{1,2,3 \text{ and } 4}$.

In the next section we discuss the general AC problem and analyse a solution.

2.7.3 A general purpose AC combining network and current source

If we consider the T-network of figure 2.33, where a and b are connected by an ideal transformer of infinite input impedance and zero output impedance, the current i flowing into a third node c is

$$i = [\mathcal{E}_b + \varrho(\mathcal{E}_a - \mathcal{E}_b) - \mathcal{E}_c] Y.$$

Consider the network of figure 2.34:

$$i' = (\mathcal{E}_a - \mathcal{E}_c)\varrho\, Y + (\mathcal{E}_b - \mathcal{E}_c)(1 - \varrho)\, Y$$

$$= i \text{ for all values of } \varrho.$$

Therefore the two networks are equivalent.

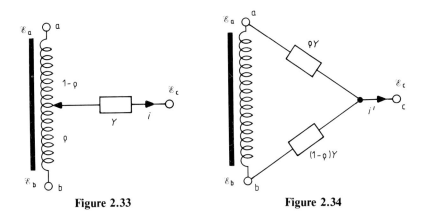

Figure 2.33   Figure 2.34

Note that if a or b are tappings on the transformer, values of $\varrho < 0$ and $\varrho > 1$ can be obtained. Therefore if $\varrho < 0$ the admittance $\varrho\, Y$ between a and c is negative, and could be either a negative conductance or a negative susceptance. The admittance between b and c would then be greater than Y.

If we take two such networks and connect them in parallel, as in figure 2.35, then the resulting network will be equivalent to that of figure 2.36 and

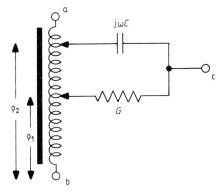

Figure 2.35

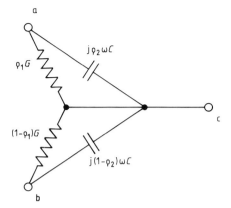

Figure 2.36

since ϱ_1 and ϱ_2 are fully adjustable, the arrangement is a general combining network with ratios of admittances of phases in quadrature to eliminate the effect on the rest of the network of any unwanted potential difference between a and b.

A sufficiently close approach to an ideal transformer can easily be obtained in practice, and sufficiently large admittances G and $j\omega C$ used so that there is no significant effect on the noise and sensitivity of the network as a whole in which the combining network is used. We give several examples of its use in the networks to be described in chapter 6.

References

Cutkosky R D 1964 Four-terminal-pair networks as precision admittance and impedance standards *Trans. IEEE Commun. Electron.* **83** 19–22

Shields J Q 1974 Measurement of four-pair admittances with two-pair bridges *Trans. IEEE Instrum. Meas.* **23** 345–52

Thompson A M 1964 AC bridge methods for the measurement of three-terminal admittances *Trans. IEEE Instrum. Meas.* **13** 189–97

——1968 An absolute determination of resistance based on a calculable standard of capacitance *Metrologia* **4** 1–7

Warshawsky I 1955 Multiple bridge circuits for measurement of small changes in resistance *Rev. Sci. Instrum.* **26** 711–15

3 Impedance Standards

In chapter 2 we discussed the electrical definition of a standard in some detail, but to make accurate measurements it is also necessary to construct standards whose value possesses adequate stability. Therefore, in this chapter we discuss the physical construction of suitable standards, the electrical properties which result from a given construction, and the principal limitations of their accuracy.

3.1 The calculable Thompson−Lampard cross-capacitor

A calculable impedance standard is one whose impedance can be computed from a knowledge of its dimensions. Its impedance is therefore known in SI units, and from it all SI impedance units can be derived. The Thompson−Lampard type of cross-capacitor has become the universally used calculable standard, having superseded calculable mutual inductors for this purpose because

(i) conducting electrodes are impermeable to electric fields and hence define the field boundaries and exclude unwanted external fields. A precise boundary for the magnetic field of an inductor is not possible without superconducting technology.

(ii) A geometry has been devised for the electrodes which has the desirable attribute that only one accurate length measurement need be made provided that straight electrodes have been aligned parallel to each other.

3.1.1 The geometrical arrangement of the electrodes—the Thompson−Lampard theorem

Dimensionally, capacitance is the product of a length, a shape factor and the permittivity of the medium. Thompson and Lampard's achievement was to show how to avoid having to make any measurements to determine the shape factor by finding a geometrical arrangement of electrodes for which a capacitance is not dependent on the shape and size of the two cross sectional dimensions, provided these remain the same as one traverses the measured length in the third, perpendicular direction. Specifically, they showed that if

50

the intersections with a plane of four electrode surfaces form the perimeter of a two-dimensional region of arbitrary shape, if the gaps between the electrodes are negligibly small and if the shape is invariant for an infinite distance in either direction perpendicular to the plane, there exists a relationship involving the capacitances per unit length between each electrode and the non-adjacent one which is independent of the shape of the region.

If the 'cross-capacitances' are C_1 and C_2 between the two opposite pairs of electrodes per unit length of the infinite system in vacuum (figure 3.1), this relationship is

$$\exp(-\pi C_1/\epsilon_0) + \exp(-\pi C_2/\epsilon_0) = 1$$

where ϵ_0 is the permittivity of vacuum. Practical instruments are invariably evacuated to avoid knowing the relative permittivity of any medium between the electrodes. The relationship is not altered by the presence of thin, longitudinally uniform layers of dielectric material on the electrode surfaces—a point of considerable practical importance since oxide layers and contaminating coatings on polished metal surfaces cannot entirely be avoided.

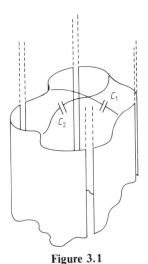

Figure 3.1

It is usual to make the arrangements of the four electrodes as symmetrical as possible so that $C_1 - C_2 = \Delta C$, where ΔC is small. In that case, the above relationship can be approximated as a series for their mean value C,

$$C = C_0[1 + \ln 2(\Delta C/C_0)^2/8 - \cdot \cdot \cdot]$$
$$\approx C_0[1 + 0.0866(\Delta C/C_0)^2]$$

where C_0 is the value obtained when $\Delta C = 0$, that is,

$$C_0 = (\epsilon_0 \ln 2)/\pi \ \mathrm{F\,m^{-1}}$$

$$= 1.953\,549\,043 \ \mathrm{pF\,m^{-1}}$$

assuming the defined value $\mu_0 = 4\pi \times 10^{-7}$ and the defined value of $c = (\epsilon_0 \mu_0)^{-1/2}$ is $299\,792\,458$ m s^{-1}. In practice, it is not difficult to adjust the cross-capacitances C_1 and C_2 to be equal to better than 1 in 10^4 (that is $\Delta C/C_0 < 10^{-4}$) so that, to an accuracy of a part in 10^9 the above numerical value of C_0 can be used.

The electrodes have, in all existing examples, been formed as circular cylinders so that the operative surfaces are the inward-facing arcs shown in figure 3.2, because such cylinders are the easiest mechanical objects to make with the necessary precision.

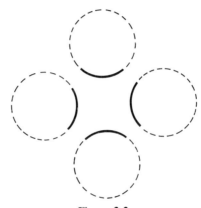

Figure 3.2

In simpler models capable of about one part per million uncertainty the electrodes are end standards of accurately known length, straight and of uniform circular cross section. The capacitor is continued on beyond these a short distance by similar rods separated from the first ones by a thin insert of insulating material. These extensions thus serve the purpose of guards to ensure that the electric field is not distorted in the region of the ends of the actual electrodes. A stable maintained standard of capacitance which is not required to be calculable can also be made in this way. The length of the electrodes may also be changed by lengthening the structure by inserting end standards of identical section and known length to produce a calculable capacitance change.

In instruments designed to achieve greater accuracies the need for electrodes of infinite length is overcome by inserting centrally up the axis between the electrode surfaces another conducting cylinder as drawn in figure 3.3(*a*). This cylinder has a diameter which is slightly smaller than the distance between opposite electrodes, and thus effectively blocks the cross-capacitances.

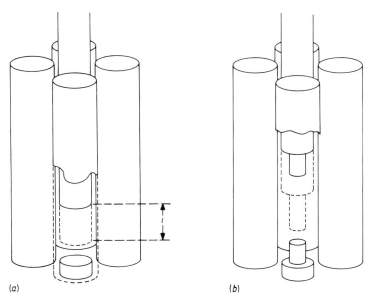

(*a*)   (*b*)

Figure 3.3

The electrostatic field does not change along the axis of the electrode system except in the region which lies just beyond the end of this inserted cylinder. Here great distortions occur but the distortion becomes negligible after a further distance down the axis of two or three times the diameter of the inserted cylinder. Therefore, since these end effects are the same with respect to the inserted cylinder, no matter what its position with respect to the fixed electrodes, they can be eliminated by inserting the cylinder a further distance *l* and observing the change produced in the cross-capacitances. It is usual, in the more precise calculable capacitors, to measure the movement *l* with a laser interferometer, the wavelength of the laser radiation in vacuum being known. *l* is often of the order of 0.5 m, corresponding to a capacitance *C* of about 1 pF. The remote end of the electrode system can be terminated in a definite manner by a similar fixed inserted cylinder as shown in figure 3.3(*b*).

The main difficulty in the practical implementation of this beautiful concept arises from the initial assumption that the electrode surfaces have straight generators all parallel to an axis so that the cross section is always the same whatever the distance along this axis. Imperfections in manufacture of the electrodes will inevitably mean that the electrode surfaces are not everywhere parallel to the axis, and consequently the non-uniform electric field at the end of the inserted cylinder will vary slightly with the position of this electrode so that the 'mechanical' distance *l* may no longer be quite identical with the 'electrical' distance. It is usual to improve this situation by providing a short extension of the cylinder of lesser diameter. For an empirically determined ratio of length to diameter of this extension the effect of a slight linear divergence of the electrodes (for example, if the axes of the electrodes are inclined slightly to one another) can be compensated so that the mechanical and electrical distances can again be made identical. The use of a properly configured extension is essential if uncertainties of less than a part in a million are to be achieved with electrodes whose departure from perfect straightness is of the order of a micrometre. The operation of this extension can be understood if it is realised that on displacing the four electrodes outwards the electric field passing round the extension increases slightly, and this compensates for the small extra amount of the fringing field intercepted by the end of the extension. It is advantageous to provide the fixed cylinder at the remote end with an identical extension because then the cross-capacitances of the assembly will be immune to sideways displacements of the electrodes from whatever cause. For example, if thermal expansion causes displacements during the time taken to traverse the movable cylinder through the distance *l* the effect of such thermal displacements will be compensated.

It is usually assumed that the electrodes are equipotential surfaces, but this is not so if they possess finite impedance. In particular, their inductance will cause errors of the order of 1 in 10^8 if it is not minimised in some way.

The type of bridge used to relate the cross-capacitances to a fixed, stable capacitor will be discussed in chapter 6.

3.1.2 Primary standards of phase angle

Calibrations of electrical power measuring equipment, of power loss in reactive components and of the phase angle of resistive components need a primary standard of phase angle. Fortunately the phase angle of a sealed two terminal-pair defined gas capacitor can be relied on to be less than 10^{-5} (a sufficient accuracy for all ordinary purposes) provided that the surfaces of the electrodes are clean and that the electric field of the observed direct capacitance does not penetrate any solid dielectric used in the mechanical construction. There is nevertheless a need for a primary standard of phase angle so that this statement can be made with confidence. There have been three main approaches to the problem.

In the past, mutual inductors at low or moderate frequencies have been used. Departure from the ideal phase angle of the mutual reactance arises from eddy current losses in the conducting materials of the windings and of any adjacent metallic matter used for mechanical support or terminations and from the effect of capacitance between the windings combined with their resistance. The energy associated with the electric field surrounding mutual inductors is small, but it also causes loss in dielectrics used for insulation etc. Nevertheless, loss effects can be kept small and a phase angle standard with an accuracy approaching a part in a million can be obtained. A mutual inductance phase angle standard is particularly useful in calibrating wattmeters for the condition of zero power factor (i.e. when the voltage and current waveforms are in quadrature).

A gas dielectric capacitor in which the spacing of the electrodes can be varied is another primary standard of phase angle. So and Shields (1979) have described the use of one for this purpose. A guard ring surrounds the low-potential electrode to ensure that no part of the electric field of the measured direct capacitance encounters solid dielectric material. A correction can be made for the loss in the resistance of the conductors between the electrodes and the terminals of the device. The only significant source of loss which cannot be directly measured and accounted for arises in dielectric films of water, metallic oxide etc on the surfaces of the electrodes. The phase defect caused by this loss, however, is inversely proportional to, and can be evaluated by varying, the total electrode separation. Stainless steel seems to be the best material with which to make the electrodes, and the construction should minimise acoustic movement resulting from the forces exerted on the electrodes by the electrostatic field. As these forces are proportional to the square of the applied voltage, this subtle cause of loss can be investigated by observing the effect of varying the voltage applied to the capacitor.

The most precise primary phase angle standard makes use of another property of the symmetrical Thompson–Lampard cross-capacitor. If one of the electrodes is coated uniformly with a lossy dielectric film, its contribution to the phase angle per unit length of the two cross-capacitances is equal and opposite. Neglecting end effects, it follows that the phase angle of the mean capacitance is independent of axially uniform, but not necessarily identical, lossy films on the electrode surfaces. The end effects can be eliminated by inserting a shielding fifth electrode, as in the usual use of the device as a standard of capacitance, and taking the difference of measurements on further insertion. Shields has adopted a slightly different approach by constructing a toroidal cross-capacitor having no ends. The cancellation of phase angles is not theoretically perfect as in the linear cross-capacitor, but is smaller than other causes of uncertainty such as acoustic loss, non-uniformity of electrode surface coating etc. A toroidal device could also be useful as a secondary maintained standard of capacitance.

Rayner (1958) has shown that the self-inductance and capacitance of an ordinary electronic component resistor, whose value is of the order of $100\ \Omega$ and which is mounted in a not too closely fitting coaxial tube, cancel. The time-constant τ $(=L/R$, where L is the effective inductance) of this two terminal-pair device is less than 10^{-9}. Therefore, although it does not constitute a primary standard, the device makes a very simple, practical phase angle standard of a few parts per million accuracy at frequencies below a few kHz.

Once a primary standard of phase angle has been established it can be used to calibrate the phase angle of other standards by means of a bridge similar to those of §§6.1.2–5. A voltage transformer or IVD may be calibrated by one of the methods of §4.4. It may be expected to have a phase angle associated with its voltage ratio of not more than a few parts in a million.

3.2 Secondary maintained standards of capacitance_____

In chapter 2 we have dealt with the question of the electrical definition of a device by defining the conditions at its terminals; here we are principally concerned with the physical form of the best impedance standards and how this form governs the influence which electrical and other environmental factors have on their apparent value. As far as electrical influences in reactances are concerned, we cannot remind ourselves too often that their value depends on the energy stored in the electric or magnetic fields associated with the standard, and anything which alters these fields will affect the value.

We will not discuss components of continuously variable impedance, that is, variable capacitors, inductors and resistors, as they do not have sufficient stability and setting accuracy to serve as impedance standards in accurate work.

3.2.1 Capacitors
Electrical energy E is stored in a capacitor in the space between and adjacent to the conductors. If the potentials of the electrodes shown in figure 3.4(a) differ by ε, $E = \varepsilon^2/2C$ where C is the capacitance. For example, for parallel plates of area A m² separated by a distance l m, which is small compared to the size of the plates, $C = \epsilon_0 \epsilon A / l$ farads. ϵ_0 is the permittivity of free space (vacuum) and ϵ is the permittivity relative to this of the dielectric between the plates.

For a stable standard it is clear that (i) a stable geometry of the conductors and (ii) a constant dielectric relative permittivity are required.

Inevitably, some part of the electric field will spill out from the capacitor, in the fringing field at the plate edges for example, or from unscreened terminals, and then any object in the vicinity of the capacitor which intercepts or modifies this field will alter the measured capacitance. Hence it is the best practice to enclose the whole capacitor and terminals in a conducting screen; the effect of the screen on the measured capacitance is

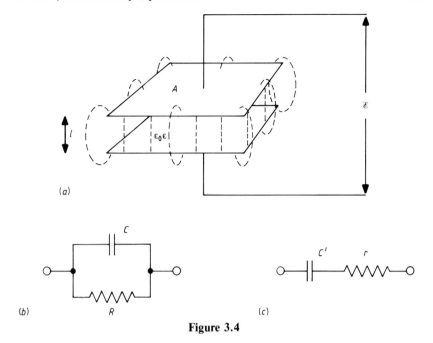

Figure 3.4

then large but definite and becomes part of the defined capacitance. Objects outside the screen no longer have any influence on the capacitance.

Two alternative lumped circuit representations of a lossy capacitor are drawn in figure 3.4(*b*), (*c*).

The capacitative and resistive parameters C, C' and R, r of these equivalent representations are connected by the relationships

$$C' = C(1 + D^2) \qquad \text{and} \qquad r = RD^2/(1 + D^2)$$

where

$$D = 1/\omega CR = \omega C'r = \tan \delta = 1/Q.$$

3.2.2 Two terminal capacitors

Capacitors made as electronic components are familiar examples of unscreened capacitors having a two terminal definition. The conducting surfaces are either two foils separated by a thin solid dielectric film, the sandwich then being rolled up tightly into a cylinder with wires connected to each foil brought out at the ends, or a stack of alternating thin conducting plates and solid dielectric with odd numbered plates being connected together to one terminal and the even to the other. In the case of the rolled construction, several connections along the length of the foil ought to be brought out and connected in parallel to minimise the inductance of the foil and limit the frequency dependence of the component. Because the electrode plates are close together, and the medium separating them has a high relative permittivity, the fringing field outside the assembly is weak compared with that inside; consequently the capacitance associated with it is small and neighbouring conducting or dielectric objects do not have any great influence on the apparent capacitance. Nevertheless, such effects are sufficiently large to exclude these components as accurate standards; they ought first to be enclosed in a screen connected to the low-potential terminal, as illustrated in figure 3.5, and precautions taken to ensure that any neighbouring objects in a measuring network are also at a low potential so that little capacitive current flows between them and the shield of the component.

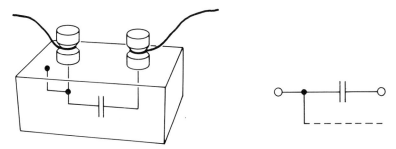

Figure 3.5 *Physical form and symbolic representation of a typical two terminal capacitor in a screening enclosure.*

The reproducibility of capacitance of this kind of standard in the presence of other nearby conductors would be about 0.1 pF, which might be acceptable in less accurate work with large value standards. As a further refinement, illustrated in figure 3.6, a removable link is often provided between the terminal attached to the high-potential electrode and a further

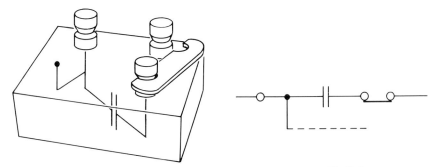

Figure 3.6 *A two terminal capacitor with a removable link.*

terminal to which the high-potential lead from the bridge or measuring instrument is connected. Removal of the link puts the small, fairly definite, capacitance between these terminals of 1 pF or less. This capacitance is in series with the standard, whose value is now defined as the change in capacitance found when the link is removed. In this way, reproducibility to 0.01 pF may be obtainable. A better construction, resulting in still greater reproducibility, is to enclose the terminals and removable link within the screen also.

3.2.3 Three terminal capacitors
A much better reproducibility is achieved by enclosing both terminals and the capacitor in an enveloping guard screen connected to the outer of one coaxial port, the other being insulated from the screen, as illustrated in figure 3.7. This guard is to be maintained at a definite low potential, but current flowing from the guard terminal is not to be included in the measurement. Notice that this is a 'three terminal' definition which is not the same as that described in §2.2.7; by completing the connection between the screen and the outer conductor of the high-potential port it can be converted to the degenerate form of a two terminal-pair described there.

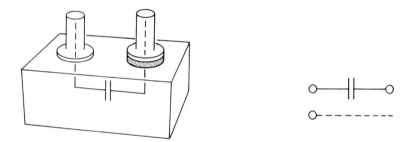

Figure 3.7 *A three terminal capacitor and its diagrammatic representation.*

3.2.4 Two and four terminal-pair capacitors
The screen is connected to the outers of all the coaxial connectors. The added complexity has the advantage already explained in §1.1 that the bridge networks into which such a standard is connected can, by current equalisation, also be made immune to external changing magnetic fields. The electrical definition of these standards is sufficiently precise for the most accurate work.

3.2.5 Types of capacitors classified by their mechanical construction
The stability of capacitors under environmental influences other than electrical ones, such as ambient temperature, humidity etc, depends on their

mechanical construction, and principally on whether a solid dielectric is used, or the plates are supported so as to be separated in dry gas or vacuum in a sealed container. The latter construction leads to capacitors of low loss whose dielectric has unchanging relative permeability, but the values of the capacitances are usually 1 nF or less, although air dielectric capacitors of up to 10 nF in value have been made. For values greater than this the large plate areas and small separations required make construction impossible and so a solid dielectric must be used to separate the electrodes. One solid dielectric, pure fused silica, has been used very successfully for low-value capacitors, having an unsurpassed stability. Unfortunately, it is not very practicable to construct large valued capacitors using this material because if fabricated into thin sheets it would be very fragile.

The principal types of capacitors used as standards with an approximate indication of some of their properties are listed in table 3.1.

The frequency dependence of the value of capacitors can be considered to arise from two causes. The first may be termed 'accidental', and is of the kind which would arise, for example, from the effect of the finite inductance of the leads coming from the electrodes of large-value capacitors. The second may be termed 'real' as it arises from the frequency dependence of the relative permittivity of the dielectric between the electrodes. This latter cause of frequency dependence is related to the dielectric loss, and a rule which is often applicable is that the order of magnitude of the proportional decrease in value of a capacitor on increasing the applied frequency ten times is the loss angle of the capacitor at the higher frequency (Rayner and Ford 1951).

Gravitational forces are an adverse environmental effect on the electrode assembly of gas or vacuum dielectric capacitors which may affect accurate work. The electrodes will deform slightly under their own weight, and this distortion will depend on their orientation so that the capacitance will change as the standard is tilted. In extreme cases it may be necessary to specify the orientation used in a calibration.

Electrode assemblies may also be prone to disturbance by vibration or accidental jarring through handling, and sudden changes in capacitance may result. This may be minimised either by making a rigid electrode structure or by ensuring that the capacitance has a stationary value, that is, it is either a maximum or a minimum with respect to electrode displacement in particular directions, or is independent of electrode displacement in others. A combination of these techniques can be used to produce a thoroughly satisfactory design. The property of stationary capacitance values or values independent of electrode displacement will ensure freedom from differential thermal expansion of the electrode structure. For independence from thermal expansion of the electrode structure as a whole, recourse must be had to either low expansion materials (Invar, fused silica etc), or to a geometrical form of the electrodes in which expansion of one part produces

Table 3.1 Properties of capacitors used as standards. The units are parts per million for the changes indicated.

Type and dielectric	Range of values	Temperature coefficient per K	10% change in relative humidity at 20°C	Barometric pressure effects per 100 mB	Stability with time per year	Loss angle at 1 kHz
(1) Sealed; dry gas	0–1 nF	1–25	—	—	1–5	1–5
(2) Unsealed; air	0–10 nF	1–25	20	60	1–5	1–25
(3) Solid; fused silica	0–100 pF	10	—	—	1	1–5
(4) Mica	1 nF–1 μF	20–50	5–25 (unsealed)	—	5–25	50–200
(5) Polystyrene	0–1 μF	–200	—	—	50	60
(6) Adjustable switched decade†	0–1 μF	Same as the components in this type of standard.				

†Types (4) or (5) switched, with a type (2) variable fine adjust. The internal resistance of the contacts gives higher values of loss angle for the higher value settings and/or higher frequencies.

an equal and opposite effect on the capacitance value to the expansion of another part. The latter, termed compensated capacitors, unfortunately are inevitably subject to transitory changes of capacitance resulting from temperature gradients within the electrode structure, and a period of several hours at a constant temperature may be required for the whole capacitor to be at the same temperature and the capacitance to be constant.

In high-accuracy work the temperature of the capacitor at the time of measurement must be recorded; in work of the highest accuracy the standard is also kept in a constant temperature environment.

Capacitors of very small value can be made by interposing between two electrodes a screen having a small aperture in it. This screen is connected to the outer case. This construction (figure 3.8) is known as a Zickner capacitor.

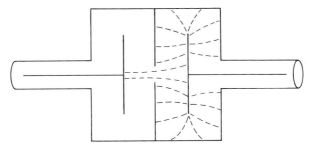

Figure 3.8 *A Zickner capacitor.*

By using removable screens with apertures of different sizes, any capacitance value from, say, 1 pF down to zero (using a screen with no aperture) can be obtained.

3.2.6 *Capacitance standards of greater than 1 μF value*
The electronics industry uses large numbers of electrolytic capacitors having values in the range from $1\,\mu$F to 1 F, principally for storing the energy of current surges at twice the line frequency (100 Hz in the UK and Europe, 120 Hz in the USA) of rectified line voltages in power supplies. Quality control and standardisation require that the capacitance of these components be measured to a modest accuracy of not better than 1 per cent. A number of commercial bridges exist which are capable of making these measurements. In order in turn to verify the accuracy of such bridges it is necessary to use stable standards of at least the decade values within this range. Electrolytic capacitors themselves do not have sufficiently stable values as a function of time, applied voltage (AC or DC) or temperature to constitute useful standards. Non-electrolytic capacitors such as polystyrene

film–metal foil rolled capacitors do possess these qualities but are only available as single components in values up to $1\,\mu\mathrm{F}$. Other plastic film materials such as polycarbonates enable similar capacitors to be made in values of up to $10\,\mu\mathrm{F}$ in a single component of reasonable physical size, and these can be paralleled to provide standards of up to, say, $100\,\mu\mathrm{F}$ in value. Whilst the stability against the various factors listed above or their phase defects are not quite as good as for the polystyrene type, they are adequate for the present purpose.

Standards having a higher capacitance value must be made as a composite component. The apparent value of a capacitor of the above type is enhanced many times by added transformers, for example, by an elegant arrangement which uses two autotransformers tapped at fractions α and β of their total turns as in figure 3.9.

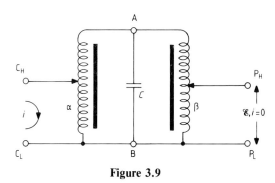

Figure 3.9

The transfer admittance Y_T of this device is the ratio of the current i flowing into and out of one pair of terminals to the open circuit EMF $\mathcal{E}$ which appears as a result at the other pair. The device is therefore inherently defined in a four terminal manner, that is, it possesses separate current and potential terminals.

The transformers are representable by equivalent T-networks, as in figure 3.10, where $Z_{1,7}$ are small impedances arising mainly from the resistance of the transformer windings with a small contribution from the transformer leakage inductances. $Z_{2,3,5\,\mathrm{and}\,6}$ are primarily composed of the transformer winding inductances which are designed to be much larger than $1/\omega^2 C$.

$$Y_T = j\omega C(1 - 1/\omega^2 LC)/\alpha\beta \qquad (3.1)$$

where $j\omega L$ is approximately the paralleled impedance of the two transformers:

$$(j\omega L)^{-1} \approx (Z_2 + Z_3)^{-1} + (Z_5 + Z_6)^{-1}. \qquad (3.2)$$

If we write

$$Y_T = j\omega C'/\alpha\beta \tag{3.3}$$

we have the very simple result that

$$C_T = C'/\alpha\beta \tag{3.4}$$

where C' is the two terminal capacitance measured between A and B with the transformers connected and the current and potential terminals open-circuited.

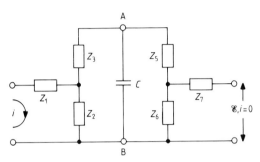

Figure 3.10

The modified capacitance C' varies slightly with voltage, because L depends on the voltage, that is, the magnetic flux density in the transformer cores. Provided that the voltage across A–B is measured and a small predetermined correction is made, this is not a serious limitation at the accuracies envisaged.

The principal disadvantage of this otherwise elegant idea is the impedance present in the upper current and potential leads. These impedances can be expressed as

$$Z_C = Z_{C_0} + \alpha(\alpha - \beta)/j\omega C' \tag{3.5}$$

and

$$Z_P = Z_{P_0} + \beta(\beta - \alpha)/j\omega C' \tag{3.6}$$

where Z_{C_0}, Z_{P_0} are the impedances measured in the C_H and P_H leads with A shorted to B.

The first term is the resistance and leakage inductance of the transformer, and the second is a positive or negative capacitance which has zero value if $\alpha = \beta$.

To obtain a ratio which is an odd power of 10 between C and C_T with this restriction requires $\alpha^{-1} = \beta^{-1} = 10^n\sqrt{10}$. Since $\sqrt{10}$ is irrational, this cannot be accomplished by using a finite number of turns on the transformers, but a close approximation can readily be obtained.

The General Radio model number 1417 standard uses this principle, and in order to provide all the decade values from $1 \, \mu\text{F}$ to 1 F in one unit it has switch-selected tapping points to the transformers. To ensure that the difference between C and C' is not too great, which is to say that the impedance of the inductance L of the transformers in parallel is much greater than that of the $1 \, \mu\text{F}$ capacitance C so that variations in the former do not cause unacceptably large variations in C_T, a large number of turns is used. This necessitates the use of a fine gauge of wire, and the two factors together result in the windings having large resistances which appear as somewhat large, mostly resistive, impedances in the C_H and P_H leads for values other than $1 \, \mu\text{F}$. These impedances are of the order of the direct impedance for the $10 \, \mu\text{F}$ range at 100 Hz, and rise to about fifty times the direct impedance for the 1 F range at 100 Hz. Note that the impedances in the C_L and P_L leads are much lower as they arise from only a short length of connecting wire, and this fact enables a standard of this kind to be compared with another using a comparatively simple bridge such as that to be described in §6.1.3.

The actual construction of the device uses networks which are associated with the switching of the transformer taps and which provide a constant small phase angle for the complete device which is independent of the value selected. By other switched networks, the same phase angle is achieved at the three frequencies of 100 Hz, 120 Hz and 1 kHz.

By making C a $100 \, \mu\text{F}$ capacitor from polycarbonate components as described above, the number of turns needed on the transformers can be reduced. For example, the transformer ratios α and β could be 127/400 turns for a 1 mF value and 40/400 turns for a 10 mF value. By having α and β equal the second terms in equations (3.5) and (3.6) above are zero, and since C is 100 times the value of that used in the General Radio design, 10 times fewer turns are needed on the transformers to give L a suitably high value. Therefore, a heavier gauge of wire can be used with a resulting lower value for the C_H and P_H lead impedances. The approximation to $\sqrt{10}$ results in a standard slightly different from the nominal value of 1 mF, but this value can be restored if required by trimming the value of C.

3.2.7 *Variation of the value of capacitors with applied voltage*
Even well designed and constructed standard capacitors are not quite perfectly linear circuit elements; their apparent value depends to some small extent on the voltage applied to them. Mechanical movement of the electrodes under the force exerted by the electric field between them, dielectric effects and effects arising from the strong field gradients at the edges of thin film electrodes coated onto solid dielectric separators are amongst the causes of this phenomenon. For the best standard capacitors whose dielectric is vacuum, dry gas, or solid pure fused silica and whose values are in the range of 0 to 1000 pF, changes are much less than a part in a

million for applied voltages in the range from 0 to 100 V commonly employed in bridge comparison networks. Nevertheless, in the most accurate work this change must be measured and accounted for, as the value of a standard must be defined as that measured at a particular applied voltage or extrapolated to a negligibly small one. It is not possible to make a reference standard having a guaranteed known small change of value with applied voltage, but the change in value of a capacitor can be determined in terms of the increments produced by factors of two increases in applied voltage, provided that three capacitors of the same nominal value are available, together with $2:-1$ and $1:-1$ ratio transformer bridges. The method described by Shields (1965) assumes that the ratio of a properly constructed $2:-1$ voltage transformer (figure 3.11) can be measured to a sufficient accuracy, say 1 in 10^9, either by the method of §§4.4.2–5 or by exchange of nominally equal capacitors in two $1:-1$ voltage bridges centred on the 0 and $\varepsilon/2$ taps.

The principle of the method is to permute three capacitors in a $2:-1$ voltage ratio bridge as shown in figure 3.11. In this way, the absolute sum of the voltage dependences is established; then, using a $1:-1$ bridge, the relative values of the voltage dependences of the capacitors when taken two at a time are found. By combining the results the voltage dependences of the individual capacitances can then be calculated.

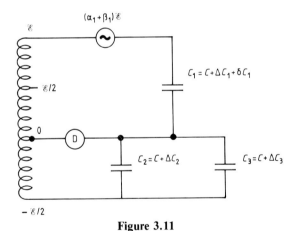

Figure 3.11

Let the increase in value of the ith capacitor ($i = 1, 2$ or 3) when the applied voltage is increased from $\varepsilon/2$ to ε be δC_i and the departure of the value of this capacitance from a nominal value C at $\varepsilon/2$ volts be ΔC_i. That is to say, at applied voltage $\varepsilon/2$ the value of the ith capacitor is

$$C_i = C + \Delta C_i$$

and at applied voltage ε is

$$C_i = C + \Delta C_i + \delta C_i.$$

The principle of the permutation bridge is shown in figure 3.11. β represents the departure of the transformer ratio from a perfect $2:-1$ ratio, and can be measured by either of the methods of §4.4. $\alpha\varepsilon$ is the small additional voltage required to balance the bridge.

The balance condition is

$$(1 + \alpha_1 + \beta)(C + \Delta C_1 + \delta C_1) = (C + \Delta C_2 + C + \Delta C_3)/2.$$

Expanding the brackets and neglecting products of small quantities gives

$$(\alpha_1 + \beta)C + \Delta C_1 + \delta C_1 = (\Delta C_2 + \Delta C_3)/2.$$

Permuting the capacitors cyclically and rebalancing gives two similar equations which, when added to the original, give

$$(\alpha_1 + \alpha_2 + \alpha_3 + 3\beta)C + (\delta C_1 + \delta C_2 + \delta C_3) = 0. \tag{3.7}$$

The capacitors are also compared by substitution in a bridge with voltage $\varepsilon/2$ applied to them, and again with voltage ε applied as shown in figure 3.12.

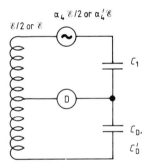

Figure 3.12

The balance conditions in these two instances are

$$(C + \Delta C_1)(1 + \alpha_4) = C_D \tag{3.8}$$

$$(C + \Delta C_1 + \delta C_1)(1 + \alpha_4') = C_D'. \tag{3.9}$$

Again neglecting products of small quantities, we obtain after subtracting equation (3.8) from equation (3.9)

$$\delta C_1 + C(\alpha_4' - \alpha_4) = C_D' - C_D. \tag{3.10}$$

Similarly

$$\delta C_2 + C(\alpha_5' - \alpha_5) = C_D' - C_D \tag{3.11}$$

and

$$\delta C_3 + C(\alpha_6' - \alpha_6) = C_D' - C_D. \tag{3.12}$$

From equations (3.7), (3.10), (3.11) and (3.12)

$$\delta C_1 = -C[(\alpha_1 + \alpha_2 + \alpha_3 + 3\beta) + 2(\alpha_4' - \alpha_4) - (\alpha_5' - \alpha_5) - (\alpha_6' - \alpha_6)]/3. \tag{3.13}$$

In this way, the change in value of a capacitor which results from reducing the applied voltage by a factor of two can be found, and therefore by making further measurements in which the voltage is halved, successive changes in capacitance $\delta C, \delta C', \delta C'' \ldots$ can also be found and a picture of the voltage dependence of a capacitor can be derived, as illustrated in figure 3.13.

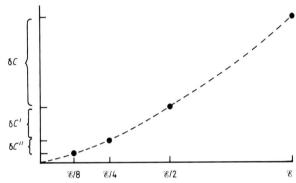

Figure 3.13 *Typical voltage dependence of the value of a capacitor.*

A two terminal-pair definition of standards is adequate for this purpose as only in-phase *changes* in their apparent relative values are required and the uncertainty produced by the uncertain contact resistance of the terminations leads only to uncertain relative quadrature errors, which are of no interest in the present instance. The loading effects of the input admittances of the standards on the transformer ratios are immaterial if these input admittances are made equal by deliberately adding more input capacitance to the lesser ones.

Once the voltage dependence of the largest value of capacitance in a decade chain of standards (for example, of the 1000 pF value) has been established, the dependence of the next lower value in the chain (100 pF) can be found directly by measurements made relative to the 1000 pF on a 10: −1 voltage ratio bridge, the voltage dependence of whose ratio has been measured (§4.4), in which the change of balance on increasing the supply voltage is observed. The higher value capacitor will need only, for the present purpose, to have its voltage dependence known to a voltage level which is 0.1 times the maximum voltage to be applied to the next lower value. The process can be repeated to establish the voltage dependence of the other capacitors in the chain.

3.3 Resistors

The primary requirement of a resistor is that its value should be as constant as possible. For high-accuracy impedance measurement the electric and magnetic fields which inevitably accompany current flow in the device should be minimised so that the reactive part of the impedance is small. Any energy loss associated with these fields should also be small so that as much as possible of the energy dissipation takes place in the resistance element. Undesirable losses include dielectric loss or loss arising from eddy currents induced in either the surrounding metal such as other parts of the resistance element (the proximity effect) and each individual part of the resistance element (the skin effect). These effects are frequency dependent and therefore lead to a frequency dependence of the resistance.

The resistance must change as little as possible with the passage of time. The principal causes of change are the state of strain of the wire and corrosion, so it is necessary to mount the resistance element with as little mechanical constraint as possible, consistent with obtaining a sound mechanical construction, and to ensure that any joints or connections between the resistive material and other conductors are made by a stable technique. Furthermore, any jointing technique such as soft soldering, crimping or pressure welding, brazing etc will produce a surrounding region of strain. The extent of this region should be minimised.

Only chemically inert substances should be used in the construction, and it is usual to immerse or embed the resistance element in oil or a flexible potting medium to exclude the atmosphere.

The most successful material for the resistance element, apart from alloys useful at cryogenic temperatures, is a four constituent or quaternary alloy of 75% Ni, 20% Cr, 2.5% Al, 2.5% Cu made under the brand name of Evanohm. The function describing its change of resistance with temperature is a parabola whose parameters can be modified to a limited but useful extent by the heat treatment given to the wire as part of the strain-relieving process during manufacture of the resistor. That is, the resistance and its temperature coefficient can be modified by the annealing process. The temperature coefficients obtainable in general are a factor of five or more lower than the older alloys of manganin etc. Like manganin, it has a low thermoelectric Peltier coefficient with respect to copper which is used for the terminals of resistors and the rest of the bridge circuit. This attribute is more important for DC applications than AC as the effects of thermal EMFS are automatically eliminated from an AC bridge balance by the repeated reversals, but even in AC applications the combination of Peltier and Thompson effects can alter the apparent resistance of a device at very low frequencies. This comes about because Peltier heating or cooling at an alloy–copper junction can combine with the thermal capacity of the junction to produce a thermoelectric EMF which persists into the next cycle of current reversal.

The main constructional techniques which attempt to put these principles into practice are described in the following paragraphs.

3.3.1 Minimally supported bare wires

This mode of construction for resistance values in the range $0.1\,\Omega$ to $10^4\,\Omega$ produces the most stable resistors of the type used for the standards of national and other laboratories. The wire used is stout enough ($0.1–3$ mm diameter is usual) to be self-supporting without touching neighbouring turns between the supporting points on the framework. The aim is to ensure that the wire is not under tension, and indeed is free to move a little so that no change in strain, and thus in resistance value, occurs. The winding on the framework should have only a small open loop area to minimise its inductance; less attention need be paid to capacitance as it is less important, particularly for the lower values of resistance. These 'S' class resistors are now designed with AC bridge use in mind, although DC methods are still the principal means of intercomparison. Binding-post terminals provide a four terminal definition which ought to be modified by adding wires to additional coaxial sockets mounted on a surrounding outer conducting can if serious AC use is intended.

The resistance element is sealed in an oil-filled metal container and the terminals are fixed to an insulating top, or through insulating bushes. A well for a thermometer descends from the top to lie alongside the resistance element. The entire container may be immersed in a thermostatted oil bath or metal block for measurement.

Figure 3.14 indicates the construction of a premier class resistance standard of this type. The resistance element is wound on a skeletal framework of insulating material. Somewhat loose winding, and careful annealing at about $180°C$ for several days ensures the vital strain-free condition.

3.3.2 Insulated (enamelled) wires wound on a flat former

This method of construction is useful for values in the range $10^3\,\Omega$ to $10^6\,\Omega$. The reactance of resistors in this range—particularly for the higher values—will arise principally from capacitance. The dielectric losses

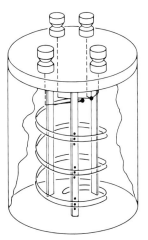

Figure 3.14

associated with capacitances should be small so that they do not affect the resistance significantly.

Attention should be given to the inductance of the wound resistor. This should also be kept small, but the fine wire, the large number of turns and their close proximity set a lower limit to what can be achieved.

Figure 3.15 shows a construction which the authors have found to give good results. Mica is usually used for the former because of its low dielectric loss when of selected quality, its dimensional stability even at the elevated temperature of around 200°C used to anneal the wound wire and its rigidity. Using a flat former minimises the cross sectional area of the wound solenoid, and hence its inductance. Using a temporary packing of PTFE shown at a, which is removed after the wire has been wound with minimum tension, ensures a just loose winding with no strain. The advance of the winding machine per turn is arranged to be slightly greater than the wire diameter so that adjacent turns are not crowded, again to avoid strain. The sharp edges of the card are protected by thin (0.05 mm) folded PTFE tape, the 'spring' of which very lightly holds the wires in place. The ends of the resistance wire are spot-welded, brazed or pressure-welded (crimped) to copper terminating wires. A short length of stouter resistance wire can be joined on to provide easier trimming of the resistance value to a tight tolerance by adjustment of its length, or by a light scraping with a very sharp tool of its surface after all annealing has been completed. (The annealing process increases the value of the completed resistor by some parts in 10^4.) Note the four terminal nature of the copper wire terminations. The completed resistance element is sealed in oil to avoid effects of the changing atmospheric humidity on the oil or mica.

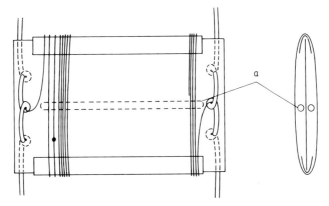

Figure 3.15 *A resistor wound on a mica card with minimum strain in the wire. The temporary* PTFE *packing pieces a are removed after winding is complete.*

3.3.3 A plane network plated on a substrate (Vishay type)

This type of resistor is useful for values in the range from $1\ \Omega$ to $10^5\ \Omega$, but this range can be extended by special techniques to $10^8\ \Omega$.

As sketched in figure 3.16, a meandering path of metal film is produced by depositing metal a few micrometres thick onto a glass substrate in a vacuum, coating it with a photoresist onto which a pattern is impressed with ultraviolet light through a mask, and then etching away the unaffected photoresist material and underlying metal. Its resistance is determined by the length, width and thickness of the meandering metal film. Wires are attached to terminating areas for connection to external terminals.

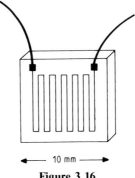

◀—— 10 mm ——▶

Figure 3.16

Low temperature coefficient alloys having a temperature coefficient as low as 1 ppm K^{-1} can be deposited, and the stability of the resulting resistor when protected by a suitable varnish and potted in a flexible medium can be remarkably good, even approaching class S standard. These resistors also have remarkably low residual inductance and capacitance, as would be expected from their small physical size and low-inductance construction. Capacitance effects are further minimised by arranging that only successive parts of the resistance path are contiguous. These resistors are therefore very useful as phase angle and AC–DC transfer standards.

This method of construction is similar to that used to make strain gauges, however, and therefore considerable care is needed in manufacture to match the thermal expansivity of the metal film and substrate and to ensure that the potting compound does not exert any permanent strain on the structure.

3.3.4 Insulated (enamelled) wire on a bobbin former

This construction, illustrated in figure 3.17, is useful for values in the range from below $1\,\Omega$ to above $10^7\,\Omega$. The bobbin is often a multisection type of low-loss ceramic material. To reduce the residual inductance of the resistor, alternate sections are wound in the opposite sense, and to reduce the residual inductance still further, the sense is also often reversed after each section has been half-wound. Nevertheless, the residual inductance almost always predominates over the residual capacitance in this type of resistor. In an improved winding method, two wires are wound together and then shorted together at one end. The two free ends are joined to the terminals of the resistor, thus making a bifilar winding of such small inductance that the residual capacitance is likely to predominate.

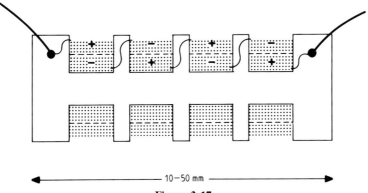

10–50 mm

Figure 3.17

3.3.5 The lumped component representation of a reactive resistor

The energy stored in the electric and magnetic fields associated with a resistor is a small fraction of the energy dissipated per cycle in the resistor. We are principally concerned with well defined resistors in which these fields lie within a surrounding outer conductor. The stored energy may be represented as in figure 3.18 as a lumped reactive component either (*a*) in parallel or (*b*) in series with the resistor.

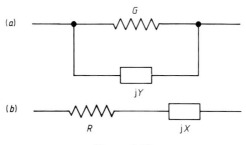

Figure 3.18

If the energy is stored principally as a magnetic field it is appropriate to represent its effect as a lumped inductive component L in series with R as in figure 3.19.

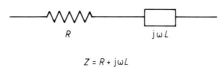

$$Z = R + j\omega L$$

Figure 3.19

Alternative possible representations are shown in figure 3.20. These equivalent circuits represent identical impedances if

$$R/(1 + j\omega CR) \equiv R' + j\omega L$$
$$R' = R/(1 + \omega^2 C^2 R^2)$$
$$L = -CR^2/(1 + \omega^2 C^2 R^2) \tag{3.14}$$
$$= -CR^2 \qquad \text{if } \omega CR \ll 1.$$

The phase angle of the resistor is the dimensionless quantity $\omega L/R'$ or $-\omega CR$. Whether a series inductance or a parallel capacitance is appropriate to represent the phase angle of a resistor usually depends on whether the phase angle is positive or negative; we will consider likely magnitudes for the actual stray capacitance and residual inductance.

The quantity $\tau = L/R = -C/R$ which has the dimensions of time, is

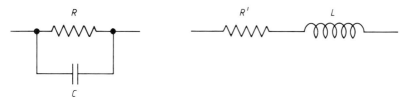

Figure 3.20

termed the time constant of the resistor. Its value should be 10^{-7} seconds or less for a well designed AC standard resistor.

The capacitance across one of the types of resistors described is likely to be of the order of a picofarad; from equation (3.14) above, for a $10^4\ \Omega$ resistor.

$$L = -10^{-12}(10^4)^{-2} = -10^{-4}\ \text{H}.$$

The actual residual inductance is unlikely to exceed 10^{-5} H so that the shunt capacitance representation is appropriate. On the other hand, for a $1\ \Omega$ resistor, the equivalent inductance for a shunt capacitance of 1 pF is

$$L = -10^{-12}(1)^2 = -10^{-12}\ \text{H}$$

and the actual residual inductance will certainly exceed this value, hence the series inductance representation is now likely to be appropriate.

The capacitive and inductive contributions can cancel, resulting in a very small time constant. As we have already noted in §3.1.2, this happens for a resistance value of the order of $100\ \Omega$, the exact value depending on the method of construction.

The representation of a resistor by three lumped components is a simplification which can depict closely the characteristics of a resistor over quite a wide range of frequencies. In a real resistor, however, the resistance, inductance and capacitance are distributed quantities, and a more precise representation can be obtained if this fact is taken into account. Additionally, these three quantities are to a certain extent frequency dependent; the values of the inductance and resistance are altered by eddy currents, and dielectric absorption may affect the capacitance values. These frequency dependent effects are not usually significant in the type of resistor over the frequency range we are concerned with in this book.

If the resistance, inductance and capacitance are uniformly distributed and have total values of R_d, L_d and C_d then the impedance is

$$Z = R_d[1 + \omega^2(2L_dC_d/3 - 2C_d^2R_d^2/15) + \cdots]$$
$$+ j\omega[L_d - C_dR_d^2/3 + \omega^2L_d(L_dC_d/3 - 2C_d^2R_d^2/5) + \cdots].$$

If the terms with ω^2 or higher powers can be neglected,

$$Z \approx R_d + j\omega(L_d - C_dR_d^2/3).$$

Where a resistor is contained within an independent electrical screen, and is in a measuring circuit such that admittances between the resistor and the screen affect the measured value, a uniformly distributed capacitance of total value K between the resistor and the screen increases the inductance by $KR_d^2/6$, ignoring terms with ω^2 or higher powers. The full expression is

$$Z = R_d[1 - \omega^2(L_dK/3 + K^2R_d^2/120) + \cdots]$$
$$+ j\omega[L + KR_d^2/6 - \omega^2L_d(L_dK/6 + K^2R_d^2/40) + \cdots].$$

The effect of an independent screen can be used to make a resistor as a whole non-reactive when in isolation it would be capacitative.

3.3.6 Auxiliary components added to resistors to reduce their phase angle
The two cases of the previous section are treated separately.

(i) The resistor has predominantly inductive residual reactance.

The lumped component representation is as in figure 3.21 and the direct impedance is

$$Z = \mathcal{E}/i = R + j\omega L.$$

If the resistor is shunted by a capacitor of value C, as in figure 3.22, C can be adjusted so that the direct impedance is now real and has the value

$$Z = R(1 + \omega^2L^2/R^2) \approx R \qquad \text{if } \omega L \ll R$$

for a value of the capacitor

$$C = L/(R^2 + \omega^2L^2) \approx L/R^2 \qquad \text{if } \omega L \ll R.$$

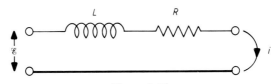

Figure 3.21

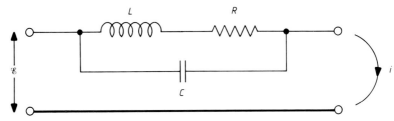

Figure 3.22

The approximations which give rise to imperfect compensation for frequencies other than the particular frequency for which the adjustment of *C* is carried out can be avoided by connecting instead a series combination of *C* and an auxiliary resistor of value equal to *R* across the original resistor, as in figure 3.23. Then $Z = R$ for all frequencies, when *C* is adjusted so that $C = L/R^2$. In practice, it is only necessary for the resistor in series with *C* to have a value approximately equal to the original to secure phase angle compensation over a wide range of frequencies.

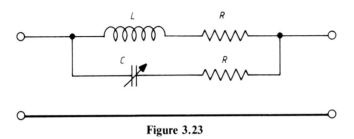

Figure 3.23

(ii) The resistor has a predominantly capacitative residual reactance.

The roles of inductor and capacitor in case (i) cannot be readily interchanged to deal with this case, for whereas an adjustable capacitor of high impedance and low phase angle can readily be obtained, this is not the case for an inductor.

A possible practical solution is to design the resistor so that it has two nominally identical sections with an available tapping point between them, each section having a capacitative phase defect. The lumped component representation is as in figure 3.24.

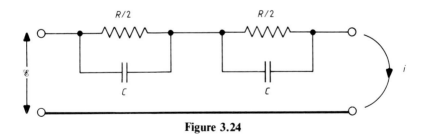

Figure 3.24

Phase angle compensation can then be carried out by connecting an adjustable capacitor C' from the tapping point to the outer conductor as in figure 3.25. A $Y-\Delta$ transformation of the network then enables it to be represented as in figure 3.26, and the direct admittance to be calculated as

$$Z = \mathcal{E}/i = 2R/(2 + j\omega CR) + j\omega C'R^2/(2 + j\omega CR)^2$$

for resistors of small phase angle, that is if $\omega CR \ll 1$ and $C' = 2C$, Z will be real and equal to R.

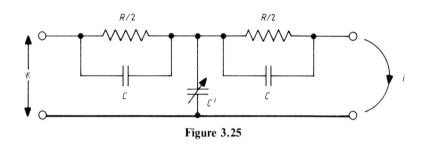

Figure 3.25

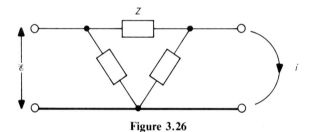

Figure 3.26

In practice, the two sections will not quite be identical, but C' can nevertheless be adjusted to make the phase angle of the transfer impedance of the device zero.

3.3.7 Variation of resistance with frequency. Frequency independent transfer standards

There are many causes of variation in the value of a resistor with frequency, and in general it is desirable that these are minimised. In particular, there is a need for resistors which, to within a desired uncertainty, have the same value for AC of frequencies up to 10 kHz as they have for DC.

Possible processes causing variation of resistance with frequency ω include:
(a) processes taking place in the actual resistance material.

(i) The skin effect, whereby alternating currents are forced to flow preferentially near the surface of a conductor to an extent dependent on ω.

(ii) The proximity effect, which is a similar alteration in the pattern of current density in a conductor caused by EMFs induced by magnetic fields from currents in adjacent conductors.

(b) processes associated with the dissipation of energy in the electric and magnetic fields associated with the current flow.

(i) Energy losses in any surrounding lossy dielectric.

(ii) The magnetic field associated with the current in the resistance element will induce circulating currents in adjacent conducting material such as the container of the resistor and mutual inductance between these and the original resistance element will include a quadrature component which will alter the apparent value of the resistor. This double mutual inductive effect is proportional to ω^2.

(c) We are concerned with the direct admittance of the device, that is, the ratio of the current flowing through a short-circuit between inner and outer at one end to the applied EMF at the other. The real part G of the direct admittance, constituting the AC value of the resistor, is modified by the distributed shunt capacitance C_d and series inductance L_d.

(i) Because of the series inductance L_d the direct impedance is modified to $Z = R + j\omega L_d$ and therefore the admittance to $(R + j\omega L_d)^{-1}$, and the real part G is modified to G' where

$$G' = G/(1 + \omega^2 L_d^2/R^2). \tag{3.15}$$

(ii) C_d modifies the apparent inductance to $L_d - C_d R_d^2/3$ (see §3.3.5) so that equation (3.15) becomes

$$G' = G/[1 + \omega^2(L_d - C_d R_d^2/3)^2/R_d^2].$$

(d) Thermoelectric processes. If the resistance element is joined to terminating conductors with which there is a large Peltier coefficient and the junction has thermal capacity, the effect of Peltier heating or cooling can cause thermoelectric EMFs to occur in the next half-cycle of low-frequency AC or reversal of DC. The resistor will appear to have a frequency dependent value at low frequencies.

Detailed explanation of these effects and calculations of their magnitudes in particular cases of interest are given by Haddad.

3.4 Mutual inductors

A current i in a primary coil produces a proportional magnetic flux which threads a secondary coil as shown in figure 3.27. That is, $\Phi = Mi$ where the constant M is termed the mutual inductance of the coil assembly. An EMF $\varepsilon = -d\Phi/dt$ is induced in the secondary coil, and is in quadrature with i, since if $i = i_0 \sin \omega t$

$$V = -i_0 \omega M \cos \omega t = -i_0 \omega M \sin (\omega t + 90°).$$

Mutual inductance has attributes which, in the past, led to mutual inductors being widely used in bridges for reactance measurements. It is an additive quantity in that if a primary coil induces magnetic fluxes in two or more secondary coils their individual induced EMFs may be added. Therefore, neglecting capacitative effects, a precise linear scale of mutual inductance can be obtained from a primary coil and tapped secondary coils.

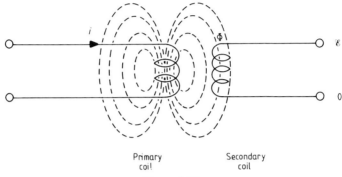

Primary   Secondary
coil   coil

Figure 3.27

An actual mutual inductor can easily be made so that it complies closely, at low frequencies of the order of 50 Hz, with the conditions for an ideal mutual inductance; that is, additiveness and negligible departure from 90° of the phase between the primary current and the secondary voltage.

Notice also the inherent four terminal nature of the device. Like a DC four terminal resistor it has separate current and potential terminals and for the definition of M to be realised in a practical measuring circuit no current must be drawn from the secondary terminals.

3.4.1 Electrical imperfections of mutual inductors
As a simple case to clarify the discussion, consider the effect of a capacitative load on the secondary of a mutual inductor, as shown in figure 3.28. We can represent the secondary (figure 3.29) as a voltage source whose magnitude is equal to the EMF induced in the unloaded secondary, and which has an inductance and a resistance:

$$i_C = \mathcal{E}' j \omega C$$

$$\mathcal{E}' = \mathcal{E} - i_C (R + j \omega L)$$

hence

$$\mathcal{E}' = \mathcal{E} / (1 - \omega^2 LC + j \omega CR).$$

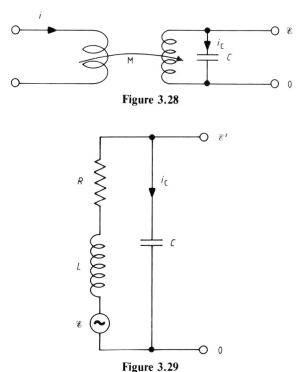

Figure 3.28

Figure 3.29

That is, there is a fractional alteration in the apparent in-phase secondary voltage equal to $\omega^2 LC$ and a fractional change in the phase angle of $90°$ equal to ωCR. In an actual case one would naturally not deliberately load the secondary in this way, but the various self-capacitances, shown in figure 3.30, between the turns of the secondary and between the secondary and primary will have a similar effect. Typical values would be $\omega = 600$ (i.e. 100 Hz), $L = 1$ H, $C = 10^{-11}$ F, $R = 10^2 \,\Omega$. Then $\mathcal{E}' = \mathcal{E}(1 + 3.6 \times 10^{-6} - \mathrm{j} \times 0.6 \times 10^{-6})$. As far as the quadrature term is concerned, other causes of loss will be more important than capacitative loading in practice.

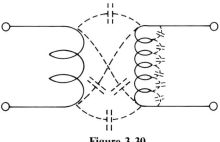

Figure 3.30

In practical bridge circuits for comparing mutual inductors it is usual to make the potentials of the primary and secondary coils definite with respect to one another by connecting one end of each together. In this case a small departure from the exact quadrature relationship between voltage and current (the phase defect) can be represented by including in the EMF across the secondary the potential drop *ir* produced across a hypothetical small resistance *r* by the primary current *i* as in the equivalent circuit of figure 3.31. This potential is in phase with *i*, in contrast to the induced EMF which, as we have seen, is in quadrature. Hence this equivalent circuit can represent the departure from an ideal 90° phase angle of a real device from whatever cause or combination of causes it may arise, as, for example, the capacitative self-loading of the device analysed above.

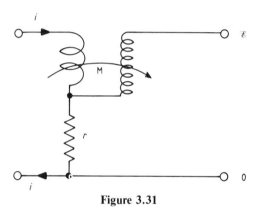

Figure 3.31

r, which is unhelpfully known as the impurity of an inductor, will in general have a very small value (a small fraction of an ohm) and is not to be confused with the very much larger resistance of the secondary, *R*, or the resistance of the primary.

Other causes contributing to *r* are eddy current losses in the windings, leakage and dielectric losses in the insulation, and any real resistance included in common with both windings when they are connected to a common measurement terminal. All these effects are frequency dependent to differing degrees and some of them contribute to frequency dependence of the mutual inductance. The phase defect of a mutual inductor represented by the equivalent circuit of figure 3.31, is $r/\omega M$.

Often in bridge work real variable resistors of small values and special construction were connected in the comparison bridge to balance out the difference in the phase defects of the mutual inductors being compared; the foregoing should give enough of a hint of the nature of mutual inductors to explain their use.

Because of their small phase defects and frequency independence at sufficiently low frequencies, mutual inductors make moderately good standard impedances, and can also be useful as 90° phase angle standards. Clearly their values are affected by causes which displace the conductors such as thermal expansion, mechanical shock etc, and by causes which influence the flux Φ, such as magnetically permeable nearby material, or sheets of conducting material (which affect the flux by eddy current opposition).

3.5 Self-inductors

Careful electrical definition of self-inductors has not yet become the general practice; they are still used as two terminal components and are therefore unsuitable for measuring in coaxial networks. Networks of twisted, individually screened wires are often used instead. It is important to realise that inductance is a property which is defined only for a complete circuit, and consequently the values of inductors must be defined as the change in inductance sensed by the measuring circuit when the configuration of the inductor is changed, usually by connecting a low-resistance link across its terminals. Some small fraction of the measuring current will flow through the inductor, the fraction depending on the ratio of the impedance of the strap plus terminal contacts to the impedance of the inductor. For low-value inductors, therefore, changes in this fraction brought about by variations of contact resistance are a problem and must be eliminated as much as possible by careful cleaning of the contacting surfaces of terminals and link. A better arrangement is to have three terminals, so that when the link is moved to short the measurement terminals, the inductor proper is open-circuited as in figure 3.32. In another arrangement, an adjacent pair of terminals is used for the current connection and another pair remote from the first for the potential connection to form a four terminal component.

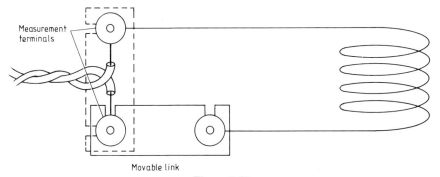

Figure 3.32

The property of inductance is not in general additive because the inductance of two inductors connected in series is modified by the mutual inductance between them. It follows that when an inductor is shorted the mutual inductance between either the shorted or unshorted configuration and the rest of the measurement circuit, particularly the leads to the measurement terminals in their vicinity, must not change. It is usual therefore to take the measurement leads, either coaxially or as a twisted pair, to a point midway between the measurement terminals of the inductor and then in opposing directions to the terminals, thus making these mutual inductances small, precise and reproducible.

Inductors not contained within a conducting box have capacitances from all parts of the coil and terminals to the surroundings including the measuring circuit, which should be contained within screens connected to its point of reference potential. The situation can then be described by the lumped parameter representation of figure 3.33. (The alternative series and parallel lumped component representations of a self-inductor and the relationships between them are given in §6.2.4.)

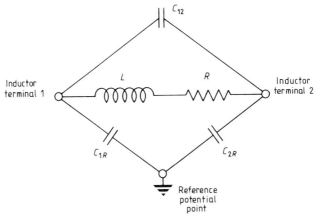

Figure 3.33

The effect of C_{1R} and C_{2R} can be eliminated with a suitable bridge configuration, but if, instead, either terminal of the inductor is connected to the bridge reference potential point, capacitance between the other terminal and the screen is added to C_{12}.

3.5.1 The frequency dependence of inductors
This may be influenced by
(i) the self-capacitance and capacitance to the surroundings and measuring circuit of the inductor

(ii) variation of eddy current effects, such as the skin effect in the conducting material of the inductor, with frequency, since eddy currents change the magnetic field distribution of the inductor. The energy loss by resistive opposition to these eddy currents will also be strongly frequency dependent, resulting in an anomalous frequency dependence of the 'Q' factor, or phase angle of the inductor.

For well designed inductors in the frequency range with which we are concerned, effect (i) will predominate, and the effective inductance $L(\omega)$ at frequency ω can be represented by the expression

$$L(\omega) = L(0)/(1 - \omega^2 L(0)C)$$

where C is the effective self-capacitance.

3.5.2 The physical construction of some commonly used self-inductance standards

There are two main types, a short solenoid or a wound toroid. A short solenoid inductor of the shape shown in figure 3.34 usually has a geometry for which the inductance is a maximum for a given length of wire. The Q of the coil is a maximum, and the effect of a slight change in one dimension of the coil on its inductance is minimised. A low temperature coefficient of inductance is achieved if the axial expansion with temperature is twice that in the radial direction.

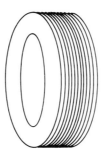

Figure 3.34

The disadvantage of the short solenoid construction is that the magnetic and electric fields from the inductor extend into the surrounding space, decreasing as $1/d^3$ at distances d large compared with the dimensions of the coil. In the vicinity of the coil where the fields are not negligible the value of the inductor will be influenced by magnetically permeable or electrically conducting materials.

The self-capacitance of the winding can be kept small—between 5 and 30 pF—by spacing the turns, and this also helps limit eddy current proximity effects, so that one can make an inductor whose Q is between 10 and 100 at 1 kHz and whose value does not change with frequency by more than 1 in 10^3 from 10 Hz to 10 kHz.

The problem of the external magnetic field is eliminated by toroidal winding if the current loop caused by the progression of the winding around the toroid is eliminated, for example, by Ayrton–Perry winding as in figure 3.35 (see §4.2.1).

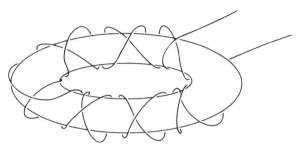

Figure 3.35

The wound inductor may be enclosed in a shielding metal can so that external electric fields are also eliminated. The resulting device is therefore well defined when provided with appropriate terminations. Unfortunately, this construction of room temperature devices (in contrast with cryogenic superconducting coils) inevitably leads to a lower Q value and higher self-capacitance with a consequent greater frequency dependence than that of the solenoid construction. Examples of toroidal inductors in use also exhibit more dependence of their inductance on ambient temperature.

References

Clothier W K 1954 A fixed gas-dielectric capacitor of high stability *Proc. IEE* **101** 453–9

—— 1965 A calculable standard capacitance *Metrologia* **1** 36–55

Cutkosky R D and Lee L H 1965 Improved ten-picofarad fused silica dielectric capacitor *J. Res. NBS* **69c** 173–9

Gibbins D L H 1963 A design for resistors of calculable AC/DC resistance ratio *Proc. IEE* **110** 335–47

Haddad R J A resistor calculable from DC to $\omega = 10^5$ rad s^{-1} *MSc. Thesis* School of Engineering and Applied Science, George Washington University

Harrison P W and Rayner G H 1967 A primary standard of mutual inductance *Metrologia* **3** 1–12

Kibble B P, Rayner G H and Swan M J 1981 Comparing capacitance standards of values greater than a microfarad *National Physical Laboratory DES Memorandum 34*

Lampard D G 1957 A new theorem in electrostatics with applications to calculable standards of capacitance *J. IEE* **104c** 271–80

Lampard D G and Cutkosky R D 1960 Some results on the cross-capacitances per unit length of cylindrical three-terminal capacitors with thin dielectric films on their electrodes *J. IEE* **107c** 112–19

Rayner G H Standard capacitors and their accuracy in practice *Notes on Applied Science No. 13* (London: HMSO)

—— 1958 The time constant of carbon composition resistors *Br. J. Appl. Phys.* **9** 240–2

Rayner G H and Ford L H 1951 The stability of mica standards of capacitance *J. Sci. Instrum.* **28** 168–71

Shields J Q 1965 Voltage dependence of precision air capacitors *J. Res. NBS* **69c** 265–74

—— 1972 Phase angle characteristics of cross-capacitors *IEEE Trans. Instrum. Meas.* **21** 365–8

So E and Shields J Q 1979 Losses in electrode surface films in gas dielectric capacitors *IEEE Trans. Instrum. Meas.* **28** 279–84

Thompson A M and Lampard D G 1956 A new theorem in electrostatics and its application to calculable standards of capacitance *Nature* **177** 888

4 Transformers

The easiest and most accurate way to compare impedance standards is in terms of the voltage or current ratio of transformer windings, because the ratio between output taps of a properly constructed transformer is not affected, at least to parts in 10^9, by aging or moderate changes in its environmental conditions (position, temperature, pressure, humidity etc). Moreover, if an accuracy of a part in a million is satisfactory, the ratio of the transformer can be taken to be equal to the ratio of the number of turns of the windings. When lower uncertainty is required, a method of calibrating the transformer ratio is needed.

In this chapter, we first discuss some general considerations which limit the performance of transformers, then we describe how to design and construct transformers to minimise these effects and finally we discuss calibration methods. This final part could be omitted on a first reading.

4.1 General considerations

A simple transformer has two or more windings threading a single ferromagnetic core which forms a magnetic circuit coupling them.

Transformers can be divided into two categories. In one, the purpose is to transmit electrical power from one circuit to another, commonly at a different voltage, and to provide isolation of one network from another, so that there is no net current flow between them. Power should be transmitted efficiently, solely by the magnetic coupling, and a high precision of voltage, or current, ratio between one winding and another is not so important a consideration. Consequently, transformers of this kind are unsuitable for measurement applications.

The other category of transformer may deliver power, but this is not in itself an important consideration. The design objective is to approach closely an ideal transformer in that the EMF induced in each and every turn should be very nearly the same and that the voltage developed by a winding should be very close to the sum of these induced EMFs, to one part in a million or less. Isolation may also be required. In the case of current transformers, the ratio of the currents in the windings is the important quantity and the accompanying voltages are only of incidental interest.

Sometimes the two windings are in fact continuous and a tapping point is provided, as in an autotransformer, and sometimes a continuous winding is divided into ten or more sections by nine or more taps.

4.1.1 The causes of departure from an ideal transformer

It is often convenient when analysing a transformer in a circuit to represent it as a combination of an ideal transformer with circuit elements which represent the imperfections in the characteristics of the actual transformer. An ideal transformer has no energy losses either in the windings or in the magnetic core, the only magnetic flux is that which threads equally all the turns of the windings and there are no capacitances within or between the windings or to the magnetic core or any surrounding screens.

In the following account of various transformers and the circuits in which they are employed, we shall be considering the causes of departure from an ideal transformer, and how they can be minimised and their effects eliminated in some circuits.

The departures of a transformer from ideal characteristics arise from several causes.

The magnetic material of the core absorbs, as well as stores, energy when it is magnetised; there is a magnetic power loss and, because magnetisation characteristics are non-linear, the power loss cannot be strictly represented by current flowing through a simple resistance. Eddy currents induced in the core material give rise to additional power loss.

The windings of a transformer will possess resistance, which at higher frequencies will be increased by eddy currents both within the conductors themselves and induced in adjacent conductors. There will also be capacitance between turns and between layers of a winding; it is shown later that these distributed capacitances can be represented by a single 'lumped' capacitance between the ends of the winding. In addition, some flux linking one winding will follow a path which does not link another winding; this flux is usually called leakage flux. The design and construction of the trans-formers used for measurement purposes aim at minimising the effects of some or all of these imperfections.

The core material should have a high stored magnetic energy compared with the energy lost per cycle. Within the restrictions imposed by other considerations, the wire for the windings should be stout to minimise the resistance and the self-inductance of the wire itself; the latter contributes to the leakage inductance. The capacitances within a winding can be reduced by spacing the individual turns and layers, but the advantages gained by this are usually outweighed by disadvantages of increased leakage flux and a higher resistance resulting from the use of finer wire for the winding. In addition it may be necessary, for example, to consider the dielectric loss of the wire insulation and other quantities of secondary importance.

However, although a transformer may depart appreciably from its ideal equivalent it will constitute a precise and stable device provided the causes of departure do not change. A cause of variability is a change in the magnetic state of the core; this is considered in the following section.

4.1.2 The magnetic core

The material used is always one of the Mumetal, or permalloy, class of nickel–iron alloys having a composition of 75% Ni and 25% Fe; the alloy with the highest permeability is known as supermumetal in the UK or supermalloy in the USA and is used where its higher cost is not an important factor. Like all magnetic alloys its magnetic properties are far from linear: the relative permeability of Mumetal alloys is shown in the graph of figure 4.1. It increases by a factor of about four from its initial value at very low flux densities up to a maximum which can be as high as 200 000 at a flux density of 0.2 T.

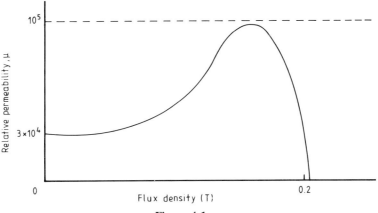

Figure 4.1

The permeability and the power loss at a given flux density change with time and depend on the past history of the material. Both mechanical and magnetic shocks are also likely to change the permeability and increase the loss, but the magnetic properties of a core can be restored to a fairly well defined state, when required, by demagnetising the material. Unfortunately, following mechanical shock or demagnetisation, the permeability decreases with time and this spontaneous change renders the demagnetised condition somewhat uncertain.

If the flux density is raised appreciably above that which produces maximum permeability, the material saturates (that is, at the peaks of an applied sinusoidal waveform, the permeability is very low). As a result the input impedance of the winding falls so that it is no longer much larger than the output impedance of a practical source, excessive voltage drop occurs in the source, the waveform is clipped, and the proportion of harmonics generated in the current and voltage waveforms associated with the transformer is increased considerably.

A solid magnetic core can be used only if the magnetic flux does not change, because when magnetic flux in the core changes, an EMF is induced in the core in planes perpendicular to the direction of the flux, and in accordance with Lenz's law, eddy currents are set up in the core such that the flux changes are opposed (figure 4.2). The EMFs, and hence the eddy currents, are proportional to the rate of change of the magnetic flux. This effect occurs in any piece of metal in a varying magnetic flux but the effect is enhanced in magnetic materials by their higher permeability. For a given uniform flux density B, the only way of reducing the magnetic flux $\int B \, dS = B \int dS$ is to reduce $\int dS$, the cross section perpendicular to B. This is usually done by building up a magnetic core from thin sheets or stampings insulated from one another, the sheets being in the direction of the flux. Since the EMF is given by $(\int B \, dS)/dt$, the larger the rate of change the smaller is $\int dS$ required to be to keep down the induced EMFs and eddy currents. That is, high frequencies require thin sheets. Unfortunately, the space factor becomes worse as the necessary insulation between adjacent strips becomes a larger proportion of the core volume.

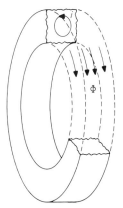

Figure 4.2 *Eddy currents induced by flux changes in a solid magnetic core.*

Eddy currents may be regarded as reducing the effective permeability of the magnetic core and increasing the total power lost via the electrical resistance of the current path during the alternating magnetisation.

Cores of the highest quality used in audio frequency applications are fabricated from high-permeability alloys formed into a strip having a thickness of 0.025 mm or 0.05 mm. The insulated strip is wound as tightly as possible into a spiral or 'clock spring' to form a toroid as sketched in figure 4.3. With thicker material of about 0.5 mm it is possible to build up a core from flat annular rings and such an assembly has the advantage that the magnetic path is truly continuous around the toroid; unfortunately it is not a practicable possibility to construct cores in this way with thinner material.

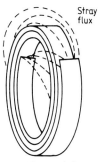

Figure 4.3

The magnetic properties of these high-permeability alloys are particularly sensitive to, and are adversely affected by, cold working, and a wound core must be annealed at high temperature in an atmosphere of hydrogen. It is then encased in a toroidal jacket filled with grease or a thick oil and sealed so that as far as possible the core is protected from mechanical stress.

When a spiral core is magnetised, the flux follows the spiral path of the strip until the outer end, where a small proportion goes back to the beginning of the spiral outside the metal. Also, another small proportion loops from outside to inside of the core on both sides of it. This looping flux is larger near the inner and outer ends of the spiral strip. Effects arising from this flux can be greatly reduced, and in fact for all practical purposes eliminated, by enclosing the core in a toroidal magnetic screen of Mumetal. By the principle of reciprocity, the toroidal flux in a magnetically shielded core can be seen to be unaffected by any magnetic fields in which the core is situated.

4.1.3 The windings; the effect of leakage inductances, capacitances and resistances

We now consider these three effects, all of which may be to some degree undesirable in a particular device, and whose individual minimisations unfortunately are in practice incompatible.

If the conductors comprising two windings could be brought into exact geometrical coincidence in space so as to link identical flux, there would be no flux which linked one winding but not the other. Since it is impossible to get two conductors to occupy the same region of space, a proportion of the flux caused by the current flowing in one winding when a load is applied to it will fail to thread the other winding and the voltage ratio of the windings will alter. This alteration can be ascribed to a small apparent internal inductance, termed the leakage inductance, of the winding.

The ratio of the induced EMFs in the windings of the transformer is $\varepsilon_1/\varepsilon_2$. Current i drawn by the load causes a flux Φ of which a small amount $\delta\Phi$ does not thread the second winding, as shown in figure 4.4.

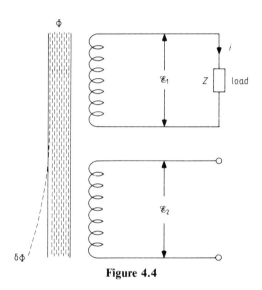

Figure 4.4

If there were no leakage flux $\delta\Phi$, voltages would be induced in the windings in proportion to their turns and the ratio $\varepsilon_1/\varepsilon_2$ would not be affected. $\delta\Phi$ gives rise to a missing EMF $-\mathrm{d}(\delta\Phi)/\mathrm{d}t = l_1\,\mathrm{d}i/\mathrm{d}t$ and modifies the ratio $\varepsilon_1/\varepsilon_2$ by this amount so that l_1 is a real physical inductance in series with the winding. The winding resistance can also be represented as a lumped resistor in series with the winding, so that we have the equivalent circuit of figure 4.5 where the windings are perfectly coupled and of zero resistance.

In a transformer where the windings are wound very closely together to minimise this effect, and spaced a greater distance away from the core or magnetic screens by insulation materials etc, the presence or absence of the core or screens will not greatly affect the magnitude of the leakage inductance because they affect only the flux which threads both windings.

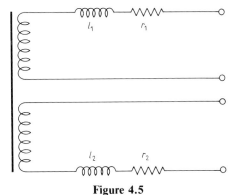

Figure 4.5

The leakage inductance and capacitance of windings also do not depend much on the core so that it is again a useful approximation in considering them to think of the device as air-cored. Any magnetic material interposed between the windings for whatever purpose would of course increase the leakage inductance enormously.

The apparent resistance of a winding is also not affected much by the presence of the core so that it is nearly the same as if the winding were air-cored. The low-frequency and DC resistance value will be augmented at higher frequencies by skin effect and proximity effect eddy currents within the conductor, particularly if the winding is closely packed or multi-layered.

In a voltage ratio transformer the potentials of successive turns of the windings increase in a regular manner until the full output voltage is reached across the complete winding. Between all pairs of turns capacitance currents will flow as shown in figure 4.6.

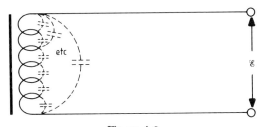

Figure 4.6

The net effect of these distributed capacitances can be represented by a single lumped capacitance. Each component capacitance δC, by virtue of transformer action, is equivalent to a capacitance $\delta C \varrho^2$ across the winding where ϱ is the fraction of the voltage across δC compared with the total winding voltage. Alternatively, a string of n capacitances ΔC in series are equivalent to a single capacitance $\Delta C/n$ across the winding. These relations are valid for all frequencies for which the phase of the current throughout the winding is constant; in practice this applies to frequencies even well above self-resonance.

Since l and r also arise from distributed phenomena, they must be shown on the winding side of this distributed shunt capacitance. Again, the winding itself is now assumed to be free of self-capacitance effects. The equivalent circuit is as in figure 4.7.

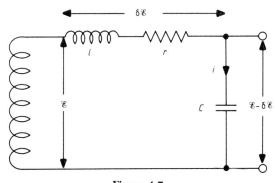

Figure 4.7

The current i drawn by the self-capacitance alters the output voltage ε of the winding to $\varepsilon - \delta\varepsilon_1$. As

$$i \approx j\omega C\varepsilon$$

$$\delta\varepsilon = i(r + j\omega l) = j\omega C\varepsilon(r + j\omega l)$$

$$= (-\omega^2 lC + j\omega Cr)\varepsilon.$$

The real part of $\delta\varepsilon$ is negative, and hence the in-phase component of voltage is increased by either the presence of self-capacitance or by an external capacitative load connected across the output terminals. l, r and C are best regarded as parameters describing the behaviour of the winding.

Cutkosky has pointed out that the deliberate introduction of a small common impedance z between two windings, as shown in figure 4.8, can be advantageous. The output impedance of one section can be decreased at the expense of another and a better overall compromise can be reached. The current drawn by a load connected between terminals A and B flows through z, thus altering the potential of terminal C so that the ratio of $\varepsilon_1/\varepsilon_2$ is immune to this loading if $z = -z_1/(\varepsilon_1/\varepsilon_2 + 1)$. In the case of a $10: -1$ ratio, $z = z_1/9$.

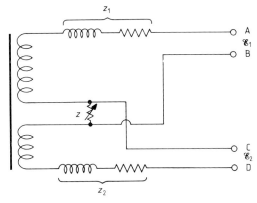

Figure 4.8

4.1.4 Representation of a non-ideal transformer. The effect of loading on its ratio

There are two kinds of transformer used in bridges. These two kinds may formally be termed two-winding and three-winding transformers, but are usually known respectively as inductive voltage dividers (or their dual, inductive current dividers) which are discussed in §§4.3.1–2 and voltage (or current) ratio transformers which are discussed in §§4.3.4–7. The essential difference between these kinds is that the energising current from a source flows through some or all of the windings of a two-winding transformer, causing small potential differences across their leakage inductances and resistances, but in a three-winding transformer the energising current flows in a primary winding which is separate from the ratio windings. The inherent accuracy of the latter arrangement is therefore greater.

An equivalent circuit of a two-winding transformer is drawn in figure 4.9.

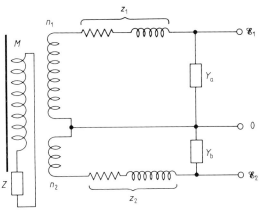

Figure 4.9

A source may be connected across either winding, or, if the windings have a common connection as shown it may be connected across both. The winding and load Z drawn to the immediate right of the core represent not an actual winding, but the extra impedance presented to the driven winding because of loss mechanisms. $z_{1,2}$ represent the series resistances and leakage inductances and $Y_{a,b}$ the internal shunt admittances of the windings. An analysis, neglecting product and higher terms, of the small quantities $z_{1,2}$ yields

$$\mathcal{E}_1/\mathcal{E}_2 = \varrho[1 + (z_1 - \varrho z_2)/(1 + \varrho)Z + (Y_b - \varrho Y_a)(z_1 + \varrho^2 z_2)/\varrho(1 + \varrho)] \quad (4.1)$$

where $\varrho = n_1/n_2$, and shows how these immittances cause the voltage ratio $\mathcal{E}_1/\mathcal{E}_2$ to depart from the nominal ratio of the windings.

The equivalent circuit of a three-winding transformer (which has a separate energising winding) when measuring the ratio of admittances $Y_{1,2}$ is shown in figure 4.10. It can be used to examine the effect that loading produces on the transformer ratio. $z_{1,2}$ are typically $10\,\mu\text{H}$ in series with $0.1\,\Omega$, so that when comparing admittances $Y_{1,2}$ of $10^{-3}\,\text{S}$ at an angular frequency of $10^4\,\text{rad s}^{-1}$, the corrections to the ratio $\mathcal{E}_1/\mathcal{E}_2$ contained in the square brackets of the first expression of equation (4.2) are of the order of 10^{-4} and must be accounted for or otherwise eliminated in accurate work.

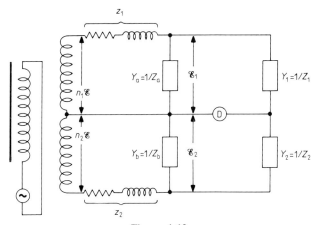

Figure 4.10

If the flux induced in the core by the primary winding in turn induces an EMF $\mathcal{E}$ per turn in the secondary windings, the ratio of EMFS $\mathcal{E}_1$ and $\mathcal{E}_2$ applied to the admittances Y_1 and Y_2 to be compared is

$$\mathcal{E}_1/\mathcal{E}_2 = (n_1\mathcal{E}/n_2\mathcal{E})[1 + z_2(Y_2 + Y_b)]/[1 + z_1(Y_1 + Y_a)]$$
$$= Y_2/Y_1 = Z_1/Z_2 \quad (4.2)$$

or $Z_1(1 + z_1/Z_a) + z_1 = \varrho[Z_2(1 + z_2/Z_b) + z_2]$.

Similar analyses have been given by Thompson (1958) and by Cutkosky and Shields (1960).

The error in the voltage ratio of a transformer caused by the loading of the shunt admittance presented by the measuring circuit to its output ports can readily be found and accounted for by measuring this shunt admittance, increasing it temporarily with a shunt load by a known ratio and extrapolating back to the zero load condition. The internal impedances $z_{1,2}$ of the transformer winding which cause the loading errors can also be calculated, using equation (4.2).

4.1.5 *The two-stage principle*

If an only approximately correct flux in a core has been created by a primary or magnetising winding, the deficiency can be made up by a much smaller flux in a second core. This latter flux is produced by a winding energised by the EMF due to the discrepancy between the exact flux required and the smaller flux provided by the magnetising winding. As applied to voltage transformers, whose equivalent circuit is shown in figure 4.11, it is desired

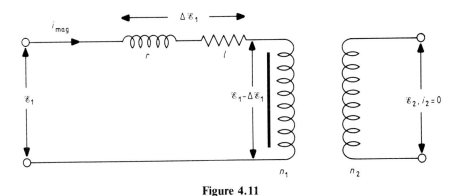

Figure 4.11

that the total flux should correspond exactly to the source EMF $\mathcal{E}_1$ defined at the input terminals, but in the ordinary single-stage design the flux corresponds to an EMF smaller by an amount $\Delta\mathcal{E}_1$ because of the voltage drop the primary current suffers in flowing through the impedance associated with the primary windings. This impedance is composed of the resistance r of the primary winding plus the leakage inductance l between primary and secondary windings.

Hence this EMF $\mathcal{E}_1 - \Delta\mathcal{E}_1$ is equal to the rate of change of flux in the transformer core,

$$\Delta\mathcal{E}_1 = i_{\mathrm{mag}}(r + \mathrm{j}\omega l)$$

and the EMF appearing across the output terminals is

$$\mathcal{E}_2 = (\mathcal{E}_1 - \Delta\mathcal{E}_1)n_2/n_1$$

instead of the ideal value $\mathcal{E}_2 = \mathcal{E}_1 n_2/n$.

A second core added and wound as shown in figure 4.12 is driven by the lost EMF $\Delta\mathcal{E}$ which, to a good approximation, can be transformed to $\Delta\mathcal{E} n_2/n_1$ and added to the EMF in the secondary winding to give the corrected voltage

$$(\mathcal{E}_1 - \Delta\mathcal{E})n_2/n_1 + \Delta\mathcal{E}_1 n_2/n_1$$

which equals $\mathcal{E}_1 n_2/n_1$ as was required. i_2 is small because of the high impedance of the inductance of the second core.

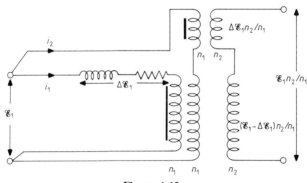

Figure 4.12

An approximate analysis of the device can be made as follows, with reference to figure 4.13.

$$i_2 = \Delta\mathcal{E}/(z_2 + z_3 + Z_3) = i_1 z_1/(z_2 + z_3 + Z_3)$$

$$= \mathcal{E}_1 z_1/(z_1 + Z_1)(z_2 + z_3 + Z_3)$$

where $Z_{1,3}$ are the impedances of the wound first and second cores, and z_{1-5} are composed of the resistances and leakage inductances of the five windings:

$$\mathcal{E}_2 = (\mathcal{E}_1 - \Delta\mathcal{E})n_2/n_1 + [\Delta\mathcal{E} - i_2(z_2 + z_3)]n_2/n_1$$

$$= \mathcal{E}_1[1 - z_1(z_2 + z_3)/(z_1 + Z_1)(z_2 + z_3 + Z_3)]n_2/n_1.$$

Hence, since $z_1 \ll Z_1$ and $z_2 + z_3 \ll Z_3$ the error term is approximately the product of two terms, z_1/Z_1 and $(z_2 + z_3)/Z_3$.

In practice, a separate auxiliary transformer can be avoided by the construction shown in figure 4.14. Remember that a winding threads *all* cores and their windings drawn to the left of it, so that winding ① is wound

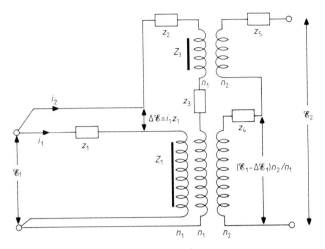

Figure 4.13

only around core 1 and winding ② is wound around both cores 1 and 2. By virtue of being around core 1, winding ② senses the flux Φ_1 and behaves like the second winding on the main transformer of the previous scheme, and by virtue of also being wound around core 2 it can generate flux Φ_2 in core 2. Since the EMFS in winding ② from Φ_1 and Φ_2 are additive, we can represent the device as in the previous figure as two windings in series, each with its individual series impedance z_3 and z_2.

Similarly, winding ③ is around both cores and can be represented as the secondary windings of both main and auxiliary transformers in series with their individual impedances z_4 and z_5.

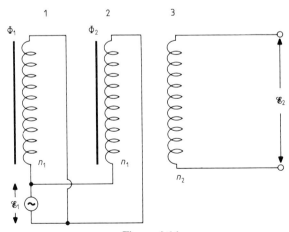

Figure 4.14

Examples of this general concept of two-staging will be discussed in the descriptions of particular devices later on in this chapter.

4.1.6 Electrical screens between windings

The capacitative current between two windings can be reduced to zero by providing two screens between them, one screen being connected to each winding. Both screens are at the same potential, but are not directly connected within the transformer as shown in figure 4.15. A screen must have an overlapped break in it so that it does not produce a shorted turn on the transformer.

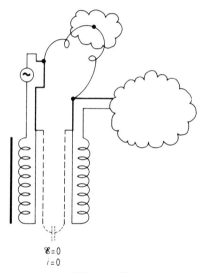

Figure 4.15

4.2 Constructional techniques

4.2.1 Design of transformer windings

A toroidal core for a transformer imposes some practical constructional limitations on the windings. The simple windings which can be put on by a toroidal winding machine do not usually produce an optimum transformer design and the best compromise for a given purpose will usually necessitate winding by hand.

In this section we describe techniques which help to ensure that a toroidal magnetic field is created by the energising winding.

Any transformer winding should be spread uniformly around the toroid so that its two ends are adjacent, as shown in figure 4.16. However, in progressing round the toroid the winding forms a loop in the plane of the

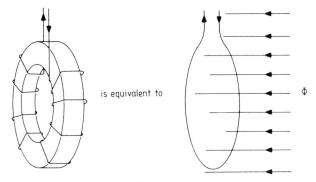

is equivalent to

Figure 4.16

toroid of a size equal to its mean diameter. The loop area may be decreased by carrying the conductor on from the end of the winding round the toroid in the opposite direction to the progression so that it finishes adjacent to the beginning of the winding. This expedient may not render the assembly immune to a field perpendicular to the toroidal plane as the core itself may be excited by such a field on account of small asymmetries in it.

A winding spread uniformly round a toroid will not respond to fields in its plane (figure 4.17), but here again asymmetries in practical cores—particularly the two ends of a clockspring construction—can make a transformer slightly susceptible to fields in some directions.

A small disadvantage of an 'anti-progression' turn as sketched in figure

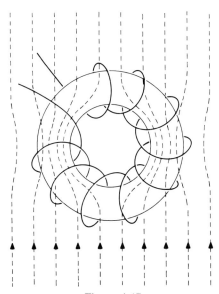

Figure 4.17

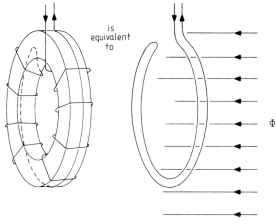

Figure 4.18

4.18 is that it does not follow the mean path of the toroid. Therefore flux in the toroidal plane may couple into the winding, or an energised transformer core may produce flux in this plane. Thus an anti-progression turn is only an approximation to a perfect solution to the problem because it evidently does not occupy the same physical volume of space as the actual winding and therefore some flux can still thread this turn and not the whole of the main winding, or vice versa.

A better solution is to use the 'Ayrton–Perry' or 'bootlace' technique as sketched in figure 4.19. Half the total number of turns required are put on the toroid in the usual way and the rest are wound on top of them, winding in

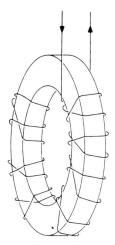

Figure 4.19

the same sense through the core but in the opposite direction around it until the beginning of the winding is reached.

A still more complex winding in which the *progression* of application of the turns is as indicated in figure 4.20 has the additional advantage of decreasing the self-capacitance of the winding considerably.

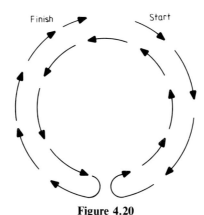

Figure 4.20

Windings of the kind we have described which do not have EMFs induced in them by external magnetic fluxes are termed 'astatic'.

In some of the following sections, the pictorial illustrations of various windings are not shown with an anti-progression feature for reasons of clarity. They should be understood to be astatically wound in practice.

4.2.2 Techniques for minimising the effect of leakage inductance, winding resistance and capacitances of ratio windings

In this section we describe winding techniques which optimise the performance of ratio windings.

The requirements are unfortunately mutually exclusive; minimum leakage inductance requires that the separate ratio windings are as close together as possible, turn by turn, to one another and this inevitably increases the self-capacitance. If the windings are made as individual successive single layers on a toroid with interposed layers of insulation, the self-capacitance of each winding will be a minimum and the capacitance between windings will also be small; the resulting capacitative currents can be further reduced by arranging that parts of the windings at similar potentials are adjacent. The effect of interwinding capacitance can be reduced to zero by appropriate screens placed between the windings as noted above.

If high accuracy of the ratio of voltage windings and small leakage inductance are required it is sometimes permissible to allow the capacitances between two windings to become large but to make them symmetrical so that they do not affect the ratio.

4.2.3 Bi-filar winding

A highly accurate ratio is achieved for the bi-filar windings of a $1:-1$ ratio transformer, which is the simplest possible ratio winding. It can give a voltage ratio that is the same as the turns ratio to a higher accuracy than any other form of winding. To achieve this, a pair of conductors, using wire from the same reel to ensure equality of resistance and insulation, is twisted together sufficiently tightly so that they lie in contact throughout their length as shown in figure 4.21. This ensures uniformity of capacitance and of leakage inductance along their length. The voltages of two completely symmetrical windings are affected identically by the presence of inductance, resistance and capacitance and their ratio is precisely equal to the ratio of the EMFs induced in the windings. Because the windings are twisted, any small differences of EMF induced in a short length of the pair by any non-toroidal stray field will sum to zero to a very close degree over the full length of the winding.

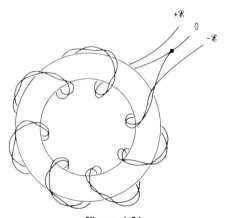

Figure 4.21

Figure 4.22 has been derived from the pictorial representation of figure 4.21 by slitting the core in the vicinity of the zero tap and opening it out to a straight line. We can now imagine the windings unravelled by pulling the $+\varepsilon$ point up and the $-\varepsilon$ point down so that the 0 points of the windings coincide, as in figure 4.23.

Near the $+\varepsilon$ and $-\varepsilon$ ends of the windings, equal capacitative loads C_1^+ and C_1^- to 0 exist. Further, capacitances symmetrical about zero occur throughout the winding so that the distributed capacitative loadings on the $+\varepsilon$ and $-\varepsilon$ sections are the same. For example, $C_2^- = C_2^+$, the voltage

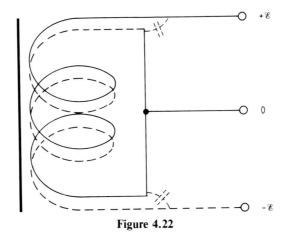

Figure 4.22

changes due to them are the same and the ratio of the output voltages remains accurately $+1: -1$.

We have assumed that the twisted pair of conductors, which is the simplest example of a 'rope' winding, is wound so that capacitances between turns are small compared to those between the twisted pair. Twisting will have an averaging effect on the small between-turns capacitances. Moreover, they are largest between those adjacent turns which have the smallest potential difference so that the capacitive current is small. Also, symmetry ensures that capacitances are equal on the two windings; that is $C_3^- = C_3^+$ etc.

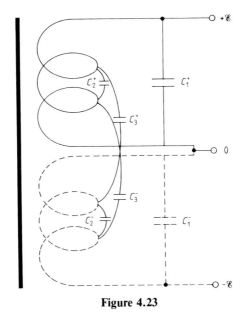

Figure 4.23

To summarise, the closeness and twisting throughout their length of the wires ensures flux either in, or escaping from, the core threads both windings equally and that the mutual leakage inductance is small. The geometrical symmetry of the windings ensures a symmetry of capacitative loading effects.

It is not surprising therefore that bi-filar winding transformers can readily be made with $+1: -1$ ratio errors of the order of 1 in 10^9 at audio frequencies where core permeabilities are high, and remain within 1 in 10^6 up to 1 MHz, provided that cores suited to this frequency are used.

Other ratios such as $+10: -1$ can be produced by suitably connecting $+1: -1$ ratios and figure 4.24 shows a possible scheme.

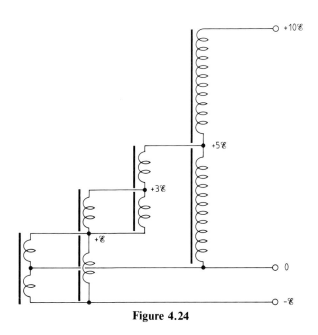

Figure 4.24

In practice, the four separate cores could well be a single core energised by a single winding from an oscillator. The successive binary windings could be shielded electrostatically and magnetically from one another.

4.2.4 Rope winding

A less perfect solution, but one which is easier to implement, is to take the conductors which are to form the ratio windings, lay them in a tight parallel bundle, twist one end to wind the complete set into a rope, and wind the complete rope a sufficient number of times (10–100) around the core, as sketched in figure 4.25. The conductors at the end of the rope are then joined

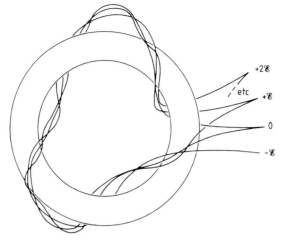

Figure 4.25

and brought to switches or connectors to form the ratio taps of the completed transformer of figure 4.26.

Thus a rope of eleven strands can be connected to form a $+10:-1$ ratio transformer. For the same reasons as with the binary winding, the conductors should be as close together as possible, and twisted, but not as in an ordinary rope where preferred strands lie in the centre and others around

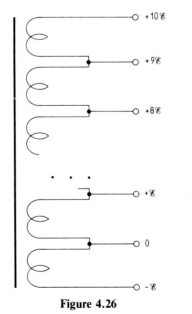

Figure 4.26

the outside. Rather, the position of the strands in the rope should be randomised by periodic interweaving of them before twisting so that interwinding capacitances and stray flux coupling are also randomised.

When the individual strands are connected in series to form a tapped winding, pairs of strands with relatively large interstrand capacitances should be connected so that there is a small voltage between them.

The departure from their nominal value of ratios produced in this way is not so small as with a binary winding (the order of 1 in 10^7 is typical), but so long as the conductors are prevented from physically altering their position in the completed transformer, and the magnetic state of the core is kept reasonably constant, the ratios are stable, and, as we shall see later, can be calibrated with an uncertainty of the order of 1 in 10^9.

4.2.5 Ribbon winding
Better equalisation of the interwinding capacitances can be achieved by winding the conductors on the core as closely as possible side by side as a ribbon with a spacing between adjacent turns of the ribbon, as indicated in figure 4.27. The ends of the wires are connected as for the rope winding.

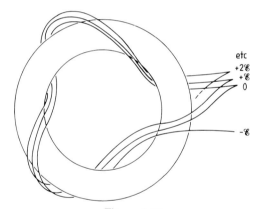

Figure 4.27

This winding method, however, has the disadvantage that the conductors are more spread out in space than with a rope winding and there are larger ratio errors from stray magnetic fields from the core (this effect can be reduced by appropriate magnetic screens) and a higher leakage inductance.

4.2.6 Strap winding
The previous windings have involved round conductors of comparatively small cross section and the internal resistance of the windings, being of the order of an ohm, is high relative to the leakage reactance. The leakage inductance will be a few microhenries. If a better compromise between

resistance and leakage inductance is required, a form of ribbon winding in which the conductors are extended outward radially from the toroid cross section to form straps can be used.

A particularly successful design for a + 10: − 1 transformer uses ten side by side straps connected in series for the 0 to + 10 sections. They are interleaved with ten further straps connected in parallel to form the 0 to − 1 section. Thus the leakage inductance is reduced and the large cross section of the straps makes their impedance low. Microhenry leakage inductances and milliohm internal resistances are achievable, justifying the rather difficult constructional technique. Figure 4.28 shows just one set of straps which forms one of the sections.

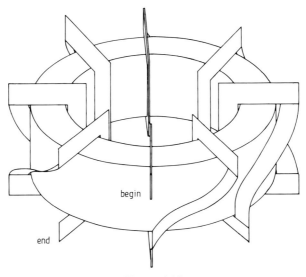

Figure 4.28

Other possible configurations are conceivable, but their advantages and disadvantages should be apparent from the above discussion.

4.2.7 *Magnetic and electric screens*
It is desirable to confine the flux produced by a winding to the interior of a toroidal space bounded by the outer surface of the winding, but a practical winding will consist of a limited number of turns of wire of finite cross section which will therefore not fill the toroidal surface uniformly. Also, toroidal flux intended to be trapped in a core will exude from the end of the 'clockspring' to go into free space a little way from the toroid and will re-enter it a little further around its circumference.

Both the effects of stray flux and capacitative currents can be reduced to negligible proportions by the use of screens of magnetic material which, as it is generally also conducting, will serve also as an electrical screen. Again, Mumetal is the best material; its mechanical formation involves sheet-metal work, preferably with the joins welded, after which the whole screen has to be heat treated again to restore its permeability. Where a de-mountable joint is necessary it must be provided with a large overlap so that the reluctance of the air gap is small enough. Soft-soldering instead of welding can be used to give electrical continuity. Figure 4.29 shows a section through a toroidal shield; connections to the wound toroid in the annular space enter through the small extension on the left. An insulating sleeve prevents a disastrous complete conducting turn threading the core.

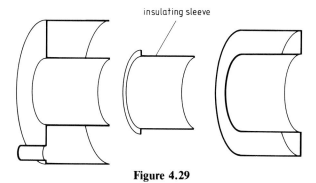

insulating sleeve

Figure 4.29

Shields can also be formed of washers and annular strips of a thin shearable form of Mumetal; although not as efficient as the above construction, the resulting shield can be sufficiently good to be useful.

Screens made simply of thick sheet copper have also been used. These rely for their magnetic effect on the skin depth phenomenon; that is to say, eddy currents induced in them tend to prevent the passage of unwanted flux through them. The skin depth at 1 kHz is about 2 mm, so this technique is only useful for higher frequencies.

Sometimes the unwanted emergence of magnetic flux is not an important effect in the device concerned, and a purely electrical screen suffices. This can be made of any conducting material and conducting paint applied to a layer of an insulating tape is frequently employed. Whatever the material used, thought must be given to the possible threading of the core by the capacitative currents to, and within, the screens.

Consider a screen around a core energised with a winding, as shown in cross section in figure 4.30.

Because the screen is a single turn around the energised core, there will be a corresponding potential difference between its energised ends, and a capacitive current will flow between them and through the screen. Therefore a small but finite current threads the core from this cause and its effect may be significant in the performance of the complete device. In the case of a magnetic screen the amount of overlap is a compromise between a near-perfect magnetic screening path and the large capacitance a big overlap would produce, permitting a relatively large capacitive current, driven by the EMF induced in the single turn of the screen across the overlap, to flow.

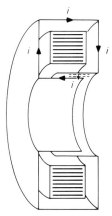

Figure 4.30 *A magnetic screen surrounding a toroidal core.*

4.2.8 Testing the goodness of a nearly toroidal field

We have emphasised that a major objective when constructing accurate transformers is to ensure as far as possible a toroidal magnetic flux by means of windings properly distributed around the core and by using surrounding toroidal magnetic screens where necessary. Then, even if the magnetic state of the core is affected by partial magnetisation or mechanical shock, the total flux still threads all ratio windings external to the toroid, and their ratio accuracy is unaffected.

It is useful to be able to test whether a nearly toroidal field has been attained, and a test is easily carried out by winding two turns in opposing senses and then connecting them to a detector as shown. T in figure 4.31 is the wound, energised toroid under test.

The junction of the loops A and C is brought together as closely as possible at B. Loop A is kept fixed with respect to the toroid whilst loop C is moved around to various parts of the circumference as required. If the excited toroid assembly has a totally toroidal field no voltage will be registered by the detector, and in so far as a voltage is detected, its ratio to that induced in a single turn around the toroid is a measure of the imperfection.

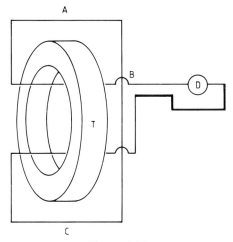

Figure 4.31

4.2.9 Connections to the output ports

The basic winding techniques for the secondary windings have already been described, but the method of bringing out the taps to the coaxial connectors requires some consideration if the highest possible accuracy is to be attained.

A good design is to mount the coaxial output connectors and the connector to the primary winding on one side of the enclosing magnetic screen, but insulated from it, and to take stout conductors from the outers, routing them alongside the tap connections to a point well within the volume of the box where they are joined together and to the zero tap as illustrated in figure 4.32. In this way, unwanted small mutual inductances between the connectors to the taps are avoided.

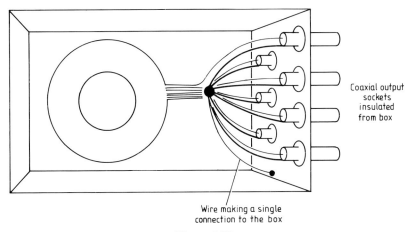

Coaxial output
sockets
insulated
from box

Wire making a single
connection to the box

Figure 4.32

The purpose of this method of mounting the output sockets is to ensure that the potentials of the outers are all well defined and small and are not affected by the flow of current returning to the transformer through the outers of other taps, as would happen if a multiplicity of current paths were provided by mounting the outer connectors directly and uninsulated on a conducting panel. The voltage of every tap is well defined, being that between the inner and outer of the coaxial output socket.

4.3 Types of transformers

The foregoing general principles have been applied to the construction of several types of transformers which have been found useful in accurate reactance bridges.

4.3.1 Inductive voltage dividers (IVDs)
These devices are of the greatest utility as they provide a voltage whose magnitude is related to the source voltage by an accurate and stable ratio which can be as finely adjusted as required, and whose phase angle, if little current is drawn, departs from that of the source by as little as one microradian.

An IVD is an autotransformer in that it has no separate primary winding. It usually has several decades to provide a fine resolution of its voltage ratio. This is accomplished in the following way.

Since we are presupposing a construction in which all the flux threads all the winding, the voltages across the turns intervals n_1 and n_2 of figure 4.33 must be such that $\mathcal{E}_2/\mathcal{E}_1 = n_2/n_1$. Hence $\mathcal{E}_2$ can be altered in finite steps by switching in different turns to alter the ratio of integers n_2/n_1.

As there is a practical limit to the number of turns of reasonable size wire

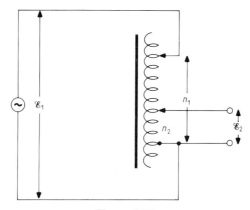

Figure 4.33

that can be used (10^2–10^3 is usual), the fineness of the steps, which is set by the ratio of numbers of complete turns, is insufficient for most applications. Hence some method of achieving what is in effect less than a complete turn is required.

Suppose another winding of m_1 turns is put on the core and connected to a winding of m_2 turns on the core of a second transformer as in figure 4.34. If the second transformer has an output winding with n_3 turns, the overall ratio of this output to the input of the first transformer is

$$(m_1/n_1)/(n_3/m_2) = (n_3/n_1)(m_1/m_2).$$

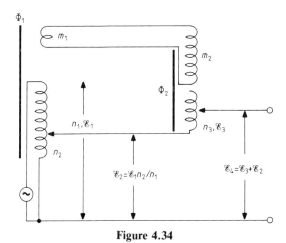

Figure 4.34

Only a small current flows because of the large inductance of the m_2 turns on the second core and the EMFS provided by the flux linkages in cores 1 and 2 are almost equal; that is, $m_1\Phi_1 = m_2\Phi_2$. If n_3 turns are wound also on core 2, the EMF induced in these turns is, to a good approximation

$$\mathcal{E}_3 = n_3\,d\Phi_2/dt = n_3(m_1/m_2)\,d\Phi_1/dt = (n_3/n_2)/(m_1/m_2)\mathcal{E}_2$$

since

$$n_2\,d\Phi_1/dt = \mathcal{E}_2 \quad \text{and} \quad \mathcal{E}_3 = (n_3/n_1)(m_1/m_2)\mathcal{E}_1.$$

Finally, the winding of n_3 turns can be put in series with the winding of n_2 turns to achieve a total output voltage

$$(n_2/n_1 + n_3m_1/n_1m_2)\mathcal{E}_1.$$

The procedure can be iterated to further cores to provide further subdivision as required.

Various modifications of this basic concept are employed to simplify construction or improve accuracy.

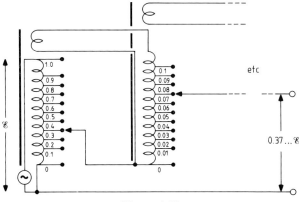

Figure 4.35

(i) The n_3 turns may be a tapped portion of the m_2 turns and the n_1 turns may also be tapped. Switch selection of taps results in the circuit illustrated in figure 4.35.

(ii) Economical use of cores can be achieved by combining each successive pair of decades onto one core, that is, putting ten sections of $10p$ turns with taps between the sections on the core together with ten sections of p turns similarly tapped. The circuit is then as shown in figure 4.36.

An IVD is mostly used as a stable AC potential or current divider, and a low output impedance is not of the utmost importance. Hence a rope method of

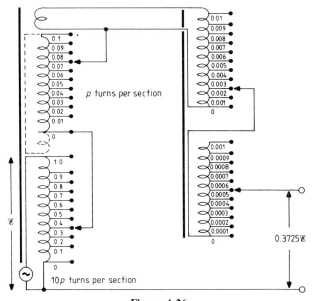

Figure 4.36

construction with a winding technique which limits the response to external flux is suitable. The first decade of the above example would consist of a rope of ten strands wound $10p$ times on the core with a 'bootlace' or non-progressive turn winding; the ends of appropriate strands are connected together at the fixed contacts of the first decade switch as shown in figure 4.37.

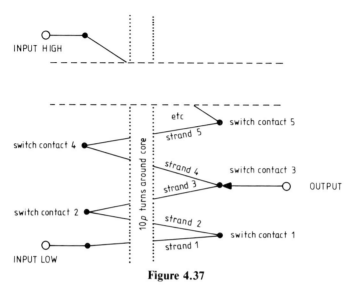

Figure 4.37

Either the ends of the strands can be labelled before the strands are twisted up into a rope, or opposite ends of the same strand can be identified with an AC continuity tester after the rope has been twisted up and wound on the core. (A DC continuity tester would leave the core in a magnetically saturated condition even if it supplied only a fraction of an ampere-turn.) Sometimes an eleventh strand is connected across the next decade switch (as shown by the dotted wire in figure 4.36); we will describe its purpose later.

The second decade of figure 4.36 is astatically wound with p turns of an eleven strand rope, 10 strands of which are connected to the second switch in a similar fashion to the connections to the first switch. The eleventh strand is used to connect across the $10q$ turns of the ten or eleven strand rope of the next transformer assembly which has also q turns of a ten-strand rope. The switches are connected as shown to form a complete four-decade divider. The construction technique can be iterated in an obvious fashion by adding further two decade cores to make a divider of $2n$ decades. Four-, six- and eight-decade dividers are commercially produced in this way.

The input impedance of the device may be increased by increasing p up to the point where the comparatively large self-capacitance of the rope winding resonates with its inductance at the frequency ω at which the divider is to be used. (A typical value would be $p = 5$ at 1 kHz.) Beyond this point the input impedance decreases again. The first core therefore has a large ratio of $\mu_r A/l$ where A is the cross sectional area, l the mean magnetic path length and μ_r its relative permeability, to maximise the inductance for a given p. Also, if B_{max} is the maximum flux in the core for which μ_r is still reasonably high just below the flux at which the core material saturates, $B_{max}A$ must be sufficient that $100pA\omega B_{max}$ exceeds the highest source voltage to be applied to the divider. Again, typically, if $A = 10^{-4}\text{ m}^2$, $B_{max} = 0.2\text{ T}$, $\omega = 10^4\text{ rad s}^{-1}$, $p = 5$, $\varepsilon_{max} = 100$ V zero to peak = 65 V RMS. If minimum quadrature component of the output voltage is desired rather than high input impedance, it is better to use the minimum number of turns required to support the intended voltage rating. The input impedance of the ratio winding may be increased in another way, by the two-stage technique described later.

No interwinding screens are necessary in these devices, but the internal capacitative loads alter the output voltage by typically not more than 1 part per million of the input at moderate frequencies, and some insulating spacing between the ropes forming successive decades on the same core is sufficient to reduce the error in the output voltage to 0.1 part per million or less of the input.

The transient conditions which occur as the switches are operated can be troublesome. It is preferable that the switches are the 'make-before-break' type, that is, contact is made with the next position before it is broken on the present position. One section of the winding is thus momentarily shorted, imposing a transient low impedance load on the source but not greatly affecting the output voltage. This is in contrast to the other possible situation of 'break-before-make' switching where the circuit to the succeeding core is momentarily open and the collapse of flux in this core produces a large momentary EMF at the output.

A better arrangement can be made, and is worthwhile for the first decades, where such effects are greatest. By means of two-pole switching with 'make-before-break' contacts on one pole and 'break-before-make' on the other a resistor of the order of 10 Ω is first switched across the taps, then the output contact is moved and finally the 10 Ω is switched out.

In the design considered above, the succeeding core is energised by a separate winding which is a strand of the p turns per section. The same arrangement is made for the second winding on a core; although it might seem unnecessary as the second winding is already correctly energised by virtue of threading the same core as the first winding, it is advantageous for two reasons. Firstly, particularly on the first core of a divider, the magnetising current flows through the windings comprising the first decade of division but, in the absence of the extra connected strand, not through the second decade windings. The voltage drop caused by this current traversing the leakage inductance and winding resistance then causes a lack of equality of potential across one section of the first decade to that across the whole of the second decade. This situation is alleviated by the arrangement referred to. Secondly, the leakage inductance which exists between the first decade rope and the second decade rope is greatly reduced by coupling them together in this way. This manifests itself in a reduced output impedance of the complete divider.

If we consider now the connection between core assemblies, then the second assembly will draw its magnetising current from the first. The resistance and leakage inductance of the energised winding on the second core will cause a loss of EMF across it because the magnetising current must flow through this impedance. Therefore the total voltage across a succeeding core assembly will not equal the voltage difference between taps on the winding preceding it. Thus small unwanted discontinuities will exist in output voltage of the complete divider as a function of switch settings.

The discontinuities can be overcome, although the accuracy is not improved, by the switching shown in figure 4.38(*a*) which arranges that succeeding windings are energised directly by the EMF between taps. The

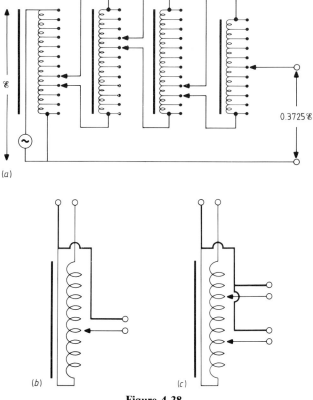

Figure 4.38

switch positions again show a voltage of 0.3725 times the input voltage being selected. As before, by winding with a rope of $10p$ turns and a rope of p turns, adjacent pairs of cores can in fact be combined as one core, with a saving of space and magnetic material.

The formal representation of an IVD we shall use is drawn in figure 4.38(b). Sometimes a single device containing some common circuitry which provides two independent outputs will be convenient, and this will be represented as in figure 4.38(c).

The two-stage principle can be applied to obtain further improvement in accuracy, as described in the next section.

4.3.2 *Two-staged IVDS*
The internal impedances of a single-decade IVD winding can be represented as in figure 4.39.

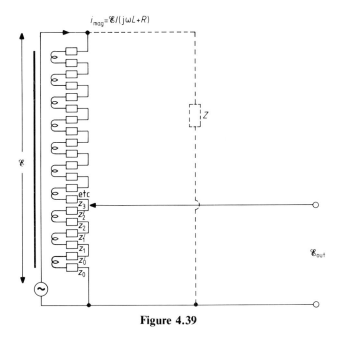

Figure 4.39

The series impedances z_q, z_q' etc arise from the resistance and leakage inductance of the individual intertap windings (the individual strands in a rope construction). They cause an additional contribution $i_{mag} \Sigma_{q=1}^{n} (z_q + z_q')$ to the EMF of the nth tap.

The input impedance of the device at the generator terminals is $Z = j\omega L + R$, where L and R represent the inductance and loss of the wound core. It is much larger than the impedances z_q, z_q', so that i_{mag} is approximately $\mathcal{E}/Z$ where $\mathcal{E}$ is the source EMF. ωL and R are likely to be about the same size.

Typically, for 50×10 turns of 10^{-3} m diameter copper wire wound on a core of 10^{-4} m² cross sectional area and 0.2 m mean magnetic path length, ωL would be about $3 \times 10^4\,\Omega$ and its resistance and leakage reactance would be about $0.1 + j10^{-2}\,\Omega$ at $\omega = 10^4$ rad s⁻¹. The wire resistance will vary by 1 per cent or more between strands and the leakage inductance by the order of 100 per cent. Several of the z_q are involved in intermediate tap voltages so the output voltage can be in error in magnitude and phase by up to a part in a million of the input voltage.

The two-stage approach can effect an immediate improvement by providing a ratio winding with a very low value of i_{mag}. The circuit for a single decade is drawn in figure 4.40.

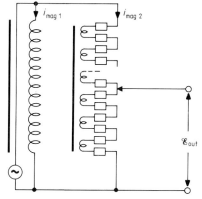

Figure 4.40 *The two-stage principle applied to an inductive voltage divider.*

The in-phase error in the output voltage at low or moderate frequencies (40–400 Hz) can be as low as a few parts in 10^9, but the quadrature voltage may be ten or more times larger.

Further decades of voltage division can be added by providing more windings and cores as before, with succeeding stages having also separate magnetising windings and two cores. Figure 4.41 is an example of a possible circuit.

The less elaborate single-stage construction would be used for further decades after the first, second and possibly third, the loss in accuracy being by then insignificant. Again, with ratio windings having $10p$ turns and p

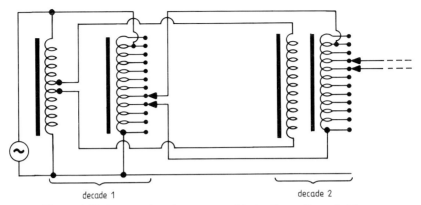

Figure 4.41 *A many-decade two-staged inductive voltage divider.*

turns, two successive decades can be combined on one core assembly needing only one magnetising winding.

Leakage inductance and interstrand capacitance will be the same for a one- or two-stage transformer and so a two-stage transformer will have similar high frequency imperfections as a single-stage transformer.

4.3.3 Injection/detection transformers
In coaxial bridges, it is very useful to be able either to introduce a generator of a small additional EMF $\Delta \varepsilon$ at a point along the inner of a coaxial line (figure 4.42), or to detect the vanishing of a current i at a point along the line (figure 4.43).

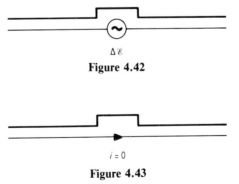

Figure 4.42

Figure 4.43

The duality principle suggests that both these objectives can be met with the same device, termed a coaxial injection or detection transformer, depending on which of the two roles it is fulfilling.

A cutaway diagram (figure 4.44) of a device described by Cutkosky will make the construction clear.

A hollow cylindrical conducting can has coaxial connectors fitted axially to its ends. Inside the can, the connectors are joined by a shielded conductor. The shield is not continuous, but has a short gap at the centre. By adding nearly touching metal discs to the shield on either side of the gap so that it is much deeper than its width the capacitance between the inner conductor through this gap to objects on the annular interior of the can is made extremely small. A high permeability toroid, bootlace wound with a large number of turns (typically 100), surrounds the coaxial line as shown, and the ends of the winding are brought out to two coaxial connectors on the can. If only a single input or detector connection is required, the other socket is shorted.

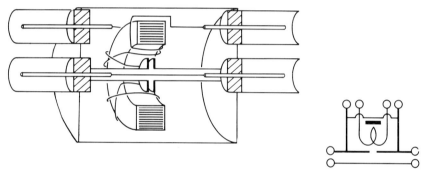

Figure 4.44 *A pictorial section and a formal representation of an injection or detection transformer.*

This construction is a n:1 transformer whose primary is the n turns on the toroid and whose secondary is the single turn of the axial conductor. The gapped screen and the can form a complete electrostatic screen which separates the primary from the secondary. The axial symmetry of the construction ensures that the device is astatic, that is, it neither emits nor responds to external magnetic flux.

A detection transformer which responds to the total current in a coaxial cable, that is, to any imbalance in the supposedly equal and opposite currents in inner and outer conductors, is also required. Figure 4.45 illustrates a suitable construction in which a high permeability toroid is wound with a hundred turns which begin and end in the inner and outer respectively of a coaxial socket. The whole winding and connections to the socket are enclosed in a toroidal shield of high permeability material which is connected to the outer of the socket. The shield has an overlapped annular gap so that it is not a shorting turn.

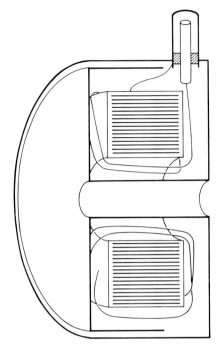

Figure 4.45 *A section through a detection transformer which responds to the unbalanced current in a coaxial cable.*

A cable to be tested for the presence of unbalanced current is threaded through the toroidal assembly, for which it is a single-turn primary, and any net current in the cable therefore causes an EMF to appear at the coaxial detector socket.

4.3.4 Use of an injection transformer as a small voltage source

If a voltage generator ε is applied to one coaxial input of the n turn secondary (typically $n = 100$) and the other coaxial socket is shorted, a voltage of ε/n is generated in the inner conductor at the centre of the device.

By virtue of the superposition theorem the use of the device can be extended to generate independent in-phase and quadrature voltages in the inner conductor. The in-phase generator ε_1 and the quadrature phase-shifting circuit whose output is represented by the generator ε_2 are arranged to have low output impedances compared with the primary impedance (typically $10^4 \, \Omega$), and are applied one to each of the two ends of the primary winding. The schematic representation of the device is then as shown in figure 4.46(*a*) with a practical circuit suitable for use at $\omega = 10^4 \, \text{rad s}^{-1}$ drawn in figure 4.46(*b*).

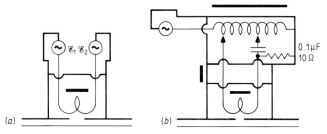

Figure 4.46

A typical value for the output impedance of the generator and phase-shifting network is 10 Ω. The generator at either input to the primary is connected to the high impedance of the primary in series with the small output impedance of the other generator so that the phase of either generator is preserved and an EMF equal to $10^{-2}(\mathscr{E}_1 + j\mathscr{E}_2)$ is generated in the coaxial inner secondary with an impedance of only $(10^{-2})^2 \times 10 = 10^{-3}\,\Omega$.

If the inductance of the secondary is insufficiently large compared with the output impedance of the generator, the two-stage technique (§4.1.5) may be employed. The circuit is then as shown in figure 4.47. A two-stage device contains two toroidal cores; the first is wound with a first-stage, or magnetising winding. The second toroid is placed alongside the wound first toroid and the second stage winding wound with the same number of turns as the first stage, but threading both cores. The pairs of coaxial inputs terminating these windings should be fed by two tracking in-phase and two quadrature generators. The generators associated with the first winding would then be corrected by the much smaller currents in the second stage

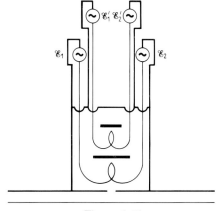

Figure 4.47

winding around both cores. The accuracy of the EMF generated in the inner can be improved by this two-stage technique from the order of 1 in 10³ to in-phase or quadrature ratio errors of 1 in 10⁶ or better.

4.3.5 Use as a detector of zero current

A current flowing down the central conductor threads the core and induces magnetic flux in it. This in turn induces an EMF in the many-turn secondary which can be sensed by a detector plugged into one of the output sockets, the other being shorted.

The precise point along the inner conductor at which the condition of zero current is sensed is defined by the position of the narrow gap in the outer, since up to this point some current can flow down the inner and across to the outer through the capacitance between them (see figure 4.48). In an equalised network, this current must return back along the outer in the same direction because as both sides of the gap are at the same potential, being shorted by the case of the device around the outside of the toroid, no capacitative current can flow across it.

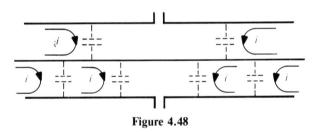

Figure 4.48

When this device is used merely to detect the presence or absence of a current the accuracy of its ratio is unimportant, and a single-stage construction is all that is necessary. The factor of n^2, where n is the number of turns on the secondary, in the impedance transfer is very valuable in matching a detector with a high input impedance into a circuit of low impedance for optimum signal-to-noise ratio.

4.3.6 Calibration of injection transformers and associated phase-change circuits

The in-phase calibration of an injection transformer A can be carried out with reference to an IVD of adequate accuracy. A circuit for a single-stage device is shown in figure 4.49. A two-stage device would need a separate feed from a tracking IVD to supply its magnetising winding.

A null of the detector D indicates zero current threading its associated detector transformer which in turn indicates equality of potential between the IVD output and that of the injection transformer A so that the IVD reading for this condition is the ratio of A.

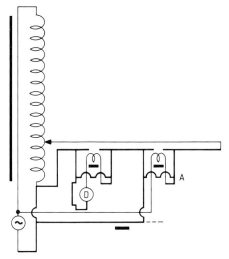

Figure 4.49

A 90° phase-change circuit and the injection transformer used with it are best thought of as a single entity to be calibrated together.

For this calibration we need components of known phase angle, and the best that are readily available are gas-filled or good quality fused silica dielectric capacitors which have phase angles of the order of 10^{-6}, and small commercial electronic resistors of about $100\,\Omega$ value. For the latter, as commonly constructed, the self-inductance is approximately balanced by their self-capacitance, and phase angles of less than 10^{-5} up to frequencies of $\omega = 10^4\,\text{rad s}^{-1}$ are usual (see §3.1.2).

These components can be assembled into a bridge to test a quadrature injection network as shown in figure 4.50.

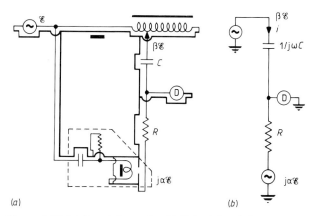

Figure 4.50 *(a) Coaxial circuit. (b) Principle of the method.*

The bridge is balanced and the detector nulled if the current through the capacitor equals that through the resistor, that is $\alpha = -\beta\omega CR$. The quadrature generator circuit is shown within the broken lines. Component values $R = 100\ \Omega$ and $C = 100$ pF are suitable for checking a phase-change network of 10^{-2} attenuation used with a 10^{-2} ratio injection transformer at a frequency of $\omega = 10^4$ rad s^{-1}.

The magnitude of the quadrature injection can be balanced by a small departure of the IVD from unity ratio, when any residual signal registered with the detector D will be due to in-phase signals from the injection system and the 100 pF and 100 Ω attenuator. That from the attenuator should not be more than 10^{-4} of the input, and by trimming the shim controls provided as part of the phase-change network of the injection system, the latter can be shimmed to agree to this accuracy.

In this discussion, only the source of quadrature voltage and its injection have been considered. It is a straightforward matter to test the complete injection system with its two IVDS, one for the in-phase voltage and one for the quadrature. The former should be set to zero when checking the quadrature voltage and vice versa.

In most applications of introducing a voltage into a network, the accuracy of the magnitude of the in-phase voltage is more important than that of the quadrature voltage. It is desirable, therefore, that any in-phase component introduced by the quadrature injection network should be trimmed to be as small as possible.

We have found that a very convenient, simple and stable way of providing trim controls for the 90° phase change circuit associated with an injection transformer is to put a few turns of enamelled copper wire wound up into a circular coil to form an inductor of value of the order of a hundred microhenries in series with the resistor whose value is so chosen that the resistance of the combination is just less than the nominal value required. The phase of the injection thus generated is adjusted to be 90° by decreasing the area of the small inductor coil by squeezing it into an elliptical form and the magnitude is adjusted by shunting C with a trimming capacitor. The inductor is best connected to the low potential end of the resistor as shown in the circuit of figure 4.51.

This method of trimming has the advantage that, if $(\omega CR)^2 \ll 1$, which is usually the case as ωCR is the attenuation introduced by the network, the circuit is trimmed correctly as to phase ($\varepsilon_{out}/\varepsilon_{in} = j\omega CR$) when $L/R = CR$. That is, the phase change is $\pi/2$ and is frequency independent. The attenuation is simply proportional to the frequency so that the same adjusted circuit can be used over a range of frequencies.

4.3.7 Voltage ratio transformers

These often form one of the major components of accurate impedance comparison bridges and are therefore very important devices. The ratio of

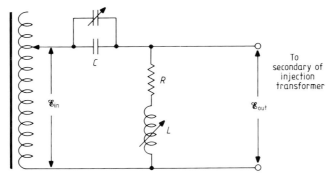

Figure 4.51

their output voltages can equal the ratio of the number of turns to within 1 part per million or so and, as we will see later, the ratio can be calibrated with even smaller uncertainty of about 1 in 10^9. Attention to detail in construction is needed, however, to ensure that this ratio is unaltered by external magnetic fluxes, is as independent as possible of the current drawn from the device, and is unaffected by changes in the mean potential of the exciting primary winding.

The core must have a large enough cross sectional area so that the total flux is sufficient for an unsaturated core to provide the required output voltage from the secondary ratio windings. It is better if the secondary windings have comparatively few turns, of the order of 100, and so to attain the voltages of the order of a hundred volts used in many bridges to measure small admittances of, for example, 10 pF, cores of large cross sectional area are required. An area of 10^{-3} m^2 is typical.

The magnetising or primary winding is best wound by one of the astatic techniques. This, together with the toroidal geometry, and the use of a toroidal magnetic screen discussed in §4.2.7, ensures a toroidal flux. The magnetic screen can also serve as one of the two electrical screens between the primary and secondary windings. By maintaining these screens at the same potential, with one connected to the primary and the other to the secondary external circuits, all capacitive currents between the screens are prevented and the secondary circuit and any measurement network of which it forms a part is completely isolated with no net current entering it via the transformer. This situation is drawn in figure 4.52.

As pointed out above, care must be taken because the screens are not equipotentials. If two screens have gaps at different places around the toroid cross section, there will be a capacitive current between them which must be provided by the networks connected to either screen. The effects of this current are nearly always undesirable, so it is best to put the gap in each screen at the same place around the cross section of the device. It is highly

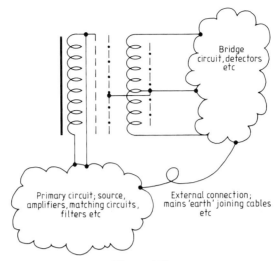

Figure 4.52

desirable to have a magnetic screen for the entire device outside the secondary ratio windings so that the external magnetic fields cannot enter and internal fields cannot escape. This screen should be connected to the secondary interwinding screen. We use a broken/dotted line as a symbol for a magnetic screen.

4.3.8 Two-stage construction
This technique can provide separate current and potential outputs to a four terminal-pair impedance and improve the definition and stability of the potential output.

The construction is as shown in figure 4.53. The winding marked * is wound only around the second core and is provided with an independent adjustable current supply. Winding and source are together called a second core excitation; their purpose is to alter the overall EMFs generated in the potential windings without altering their ratio. For example, the interwinding capacitance of the potential windings will cause capacitance currents to flow in these windings; these currents energise the second core and in turn cause the overall EMF of the potential windings to be altered, considerably increasing the in-phase component. Some opposing second core excitation is then required to restore these EMFs to equality with those produced by the current windings. The second core excitation supplies the current flowing in the interwinding capacitances. Potential windings should provide no current to external components; to minimise error arising from potential drop of any small current they deliver, winding techniques which lead to low output impedances should be used. The method described in

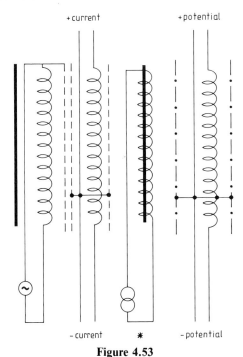

+current   +potential

−current   ✳   −potential

Figure 4.53

§4.1.3 for reducing the output impedance of one section at the expense of another to provide a better compromise may well be beneficially employed also.

4.3.9 Isolation and matching transformers

These are special cases of transformers which have only one secondary winding. Their screens are arranged as described in §4.3.7 so that there is only magnetic coupling between the two windings. An isolation transformer constructed in this way can be put between two networks to prevent capacitative currents flowing between them. Also, the impedance matching ratio $(n_1/n_2)^2$ can be used to match the output impedance of one sub-net to the input impedance of another. A common example is the use of a transformer to match the output impedance of a bridge to the input impedance of a null detector in order to obtain best sensitivity or signal-to-noise ratio, and at the same time to prevent capacitative currents entering the detector and causing an erroneous indication. Another, the dual of this situation, is matching a source output impedance to a bridge input impedance whilst again preventing unwanted capacitative currents from entering the bridge network. These points are discussed further in §5.5.

4.3.10 Current ratio transformers

The above discussion on voltage transformers can be applied to the operation of current transformers by invoking the principle of duality. That is, all sources can be altered to detectors, and detectors to sources, and current and voltage transformers interchanged throughout a circuit. The resulting representation may help those who are more familiar with voltage transformers in bridge circuits.

The action of a conventional single-stage current ratio transformer can be understood in the following way, with reference to figure 4.54. The equality and opposing sense of the ampere turns, $n_1 i_1 = n_2 i_2$, is sufficient to ensure that they together induce zero magnetic flux in the core and therefore zero EMF in the detector winding. The magnetic and electrostatic screens connected to the low potential end of the toroidal windings ensure that there is no direct coupling between current and detector windings, that is, coupling other than the magnetic coupling through the core and the space within the toroidal detector winding. In this way, an accurate and stable ratio is obtained, as in the case of the voltage ratio transformer.

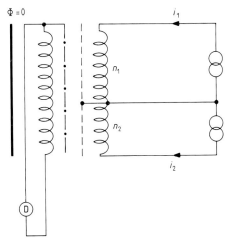

Figure 4.54 *A current ratio transformer used to compare the ratio i_1/i_2 of two currents.*

The number of turns on the detector winding and the core cross section and reluctance can be selected to produce the optimum impedance to present to the detector for the best ratio of signal-to-noise.

The finite leakage inductance and resistance of the two ratio windings in conjunction with the finite parallel output admittance of the current sources and the internal, mostly capacitive, admittances of the windings cause

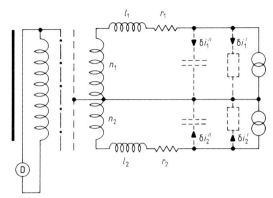

Figure 4.55

ratio errors in a fashion which is the dual of the situation in a voltage ratio transformer, and is illustrated in figure 4.55.

4.4 Calibration of transformers

All the types of transformers so far described are exceedingly accurate ratio devices within the audio frequency range, the ratios being usually within a part per million of the ratio of the number of turns comprising the windings, and the difference can itself be a stable quantity for a given device. This difference or ratio error can be experimentally measured to a high accuracy, and this measurement constitutes a calibration of the transformer. The dependence, if any, of the ratio error on the voltage applied to the transformer may be measured.

4.4.1 Calibrating voltage ratio transformers using a calibration transformer with a single output voltage

The commonest method of calibration requires that not only the taps needed to establish the desired ratio (for example, in the case of a $+10:-1$ ratio, the -1 tap, the zero tap and the $+10$ tap) but also certain others must be available. We will assume for simplicity that all twelve taps -1, 0, $+1$, $+2, \ldots, +10$ are available. The principle is illustrated in figure 4.56.

The output voltage of the secondary winding of the calibrating transformer is adjusted until it matches that between the two adjacent taps 0 and -1 of the transformer to be calibrated with the auxiliary small voltage source α_0 set to zero, that is, $\alpha_0 = 0$. Then no current flows in the loop as is shown by a null indication of the detector D.

Whilst keeping the output voltages of both transformers fixed, connections are made to successively higher pairs of output taps of the

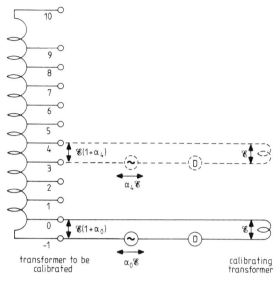

Figure 4.56 *The principle of calibrating a transformer.*

transformer to be calibrated and the detector again nulled by adjusting the auxiliary voltage source to the respective values $(\alpha_1, \alpha_2, \ldots, \alpha_{10}) \times \varepsilon$. In general, α is a complex quantity. The loop drawn in broken lines shows the connection made to determine α_4.

Thus the voltages between successive taps of the transformer to be calibrated are $\varepsilon(1 + \alpha_1)$, $\varepsilon(1 + \alpha_2)$, $\ldots$, $\varepsilon(1 + \alpha_{10})$ and the voltage between the 0 and $+ 10\varepsilon$ taps is the sum of these voltages

$$\varepsilon \left(10 + \sum_{i=1}^{10} \alpha_i \right) = 10\varepsilon \left(1 + 0.1 \sum_{i=1}^{10} \alpha_i \right).$$

The $+ 10 : -1$ voltage ratio of the transformer is

$$10\varepsilon \left(1 + 0.1 \sum_{i=1}^{10} \alpha_i \right) : -\varepsilon$$

which departs from the nominal $+ 10 : -1$ ratio by the fraction

$$0.1 \sum_{i=1}^{10} \alpha_i.$$

In the practical circuit shown in figure 4.57, the output voltage of the calibration transformer is adjusted as to its in-phase component by IVD A. A small quadrature component adjustment is provided by the injection transformer, phase-shift network and IVD A'. The adjustable series voltage $\alpha\varepsilon$ is supplied by an injection transformer, phase-shift network and two IVDs

B and B′ as described in §4.3.4. We have supposed for simplicity that the detector is an isolated battery-powered type.

This simple approach works well enough to moderate accuracies at low frequencies, but, as we have presented it, suffers from four serious drawbacks.

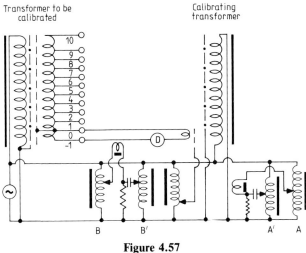

Figure 4.57

(i) If the overall potential of the surroundings is that of the zero tap, capacitive currents will flow from both transformers through all parts of the calibration loop to the surroundings and some of these currents will flow through the detector and give incorrect balances.

(ii) As drawn, the circuit is of a single conductor, open loop construction instead of a conductor-pair coaxial design. Consequently there will be many mutual interactions between the loop and other parts of the circuit and these will inject voltages in the loop, again causing wrong balances.

(iii) The use of an isolated detector may place a limit on the achievable sensitivity if its characteristics are not suited to the low impedance circuit of the loop.

(iv) The relative constancy of the output voltages of the two transformers depends on the constancy of both of their primary to secondary voltage ratios. These cannot be expected to be as stable as the voltage ratio of, for example, two secondary windings of one transformer because of the inevitable lack of spatial coincidence of the primary and secondary windings owing to the necessary interposing of shields. The amount of stray flux in this gap (which therefore threads the secondary but not the primary winding) stays constant, but the flux in the core depends on the magnetic state of the core and this magnetic state is easily altered by temperature changes, mechanical shocks, changes in the source voltage for the whole network, and the passage of time. If one of the interposed shields is a magnetic shield as well as an electrostatic one, this situation is exacerbated. The effect may well limit the calibration accuracy to 1 in 10^7.

The first three drawbacks can be overcome by appropriate arrangements of the circuit, albeit with some complications, but the fourth is fundamental to all transformers when the ratio between two windings is provided through the magnetic field in the core, and it is impossible to circumvent it directly.

4.4.2 *Calibration with a 1: − 1 ratio transformer*

This last problem requires a slight elaboration of the above principle for its solution and this, together with effective ways of eliminating the other causes of uncertainty listed, has resulted in the somewhat complex technique now to be described. If desired, some of the measures taken can be applied to the one-loop system just described.

Voltage ratio calibrations with uncertainties of the order of 1 in 10^9 are fairly readily achievable, but to realise the full accuracy requires careful attention to the topology of the circuit and its electrostatic screens.

The principle of the method is to replace the somewhat uncertain primary to secondary ratio of the calibrating transformer by the stable 1: − 1 ratio between two bi-filar secondary windings (see §4.2.3). The error in the 1: − 1 ratio can readily be found by interchanging the roles of the ratio windings in the calibration network, which now has two loops and two detectors as shown. Only the secondary windings of the transformers are shown in figure 4.58. T_A is the transformer being calibrated; three successive taps $n-1$, n, $n+1$ nominally equally separated in voltage are shown. T_C is the nominally 1: − 1 ratio calibration transformer. The complex quantities α, ϵ and a_n represent small proportional voltages.

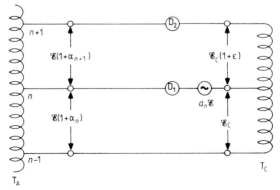

Figure 4.58

The total voltage $2\mathcal{E}_C + \epsilon \mathcal{E}_C$ developed by the $1:-1$ ratio winding of T_C is adjusted to exact equality by altering its primary voltage, until the detector D_2 shows a null indication, with the voltage $\mathcal{E}(2 + \alpha_n + \alpha_{n+1})$ between the $n-1$ and $n+1$ taps of T_A. At the same time the difference of ratios of the voltages of the two transformers is balanced and measured by the small in-phase and quadrature generator $a_n\mathcal{E}_C$ using the null detector D_1.

These balances are achieved simultaneously by a quickly converging iterative procedure. When observing D_2 the comparatively high impedance of D_1 ensures that D_2 responds to the desired condition of equal total voltage. D_1 is arranged to have a higher impedance than the whole outer loop and, whilst bringing D_1 to a null, D_2 is shorted.

It is important to realise that because of the transformer action the impedance $a_n\mathcal{E}$ of the loops to the source is just the leakage inductance and series resistance of the windings and hence large currents will flow as a result of a small lack of balance in the voltages; this effect gives the method its great sensitivity.

When both detectors are nulled we have

$$\mathcal{E}(1+\alpha_n) + \mathcal{E}(1+\alpha_{n+1}) = \mathcal{E}_C + \mathcal{E}_C(1+\epsilon) \tag{4.3}$$

for no current in D_2; this is the auxiliary balance condition which ensures that neither transformer is delivering current.

Also,

$$\mathcal{E}(1+\alpha_{n+1}+a_n)/\mathcal{E}(1+\alpha_n-a_n) = \mathcal{E}_C(1+\epsilon)/\mathcal{E}_C \tag{4.4}$$

for the main balance condition of no current in D_1. An equation similar to equation (4.4) is obtained by reversing the connections between T_C and T_A at the defining output connectors of T_A and rebalancing:

$$\mathcal{E}(1+\alpha_{n+1}+a_n')/\mathcal{E}(1+\alpha_n-a_n') = \mathcal{E}_C/\mathcal{E}_C(1+\epsilon'). \tag{4.5}$$

From equations (4.4) and (4.5), neglecting products of small quantities,

$$\epsilon + \epsilon' = 2(a_n' - a_n) \tag{4.6}$$

and

$$\alpha_{n+1} - \alpha_n + (a_n + a_n') = (\epsilon - \epsilon')/2. \tag{4.7}$$

If the calibration transformer ratio is independent of the connections to the circuit, $\epsilon = \epsilon'$ and then

$$\alpha_{n+1} - \alpha_n = -\bar{a}_n \tag{4.8}$$

where

$$\bar{a}_n = (a_n + a_n'). \tag{4.9}$$

The recurrence relation (4.8), obtained by repeated observations a_n as T_C is stepped along the taps of T_A, enables any ratio between the intertap voltages

to be found. In particular, the departure of the transformer ratio from a
nominal $+10: -1$ value is given by

$$0.1 \sum_{i=0}^{10} (10-i)\bar{a}_i.$$

However, if only the $+10: -1$ ratio is required to be measured, then one can
employ a scheme involving four instead of ten measurements, with a
consequent reduction of error. For example, one could use the three taps of
T_C to connect to, successively, the $-1, 0, +1$ taps, then the $-1, +1, +3$
taps, then $+1, +3, +5$ taps and finally the $0, +5, +10$ taps, with the
observed injected voltages $\bar{a}_1 \mathcal{E}_C$, $\bar{a}_3 \mathcal{E}_C$, $\bar{a}_5 \mathcal{E}_C$, $\bar{a}_{10} \mathcal{E}_C$ respectively.

Figure 4.59 illustrates this scheme. The departure from a true $10: -1$
ratio is

$$0.1(6\bar{a}_1 + 4\bar{a}_3 + 2\bar{a}_5 + \bar{a}_{10}).$$

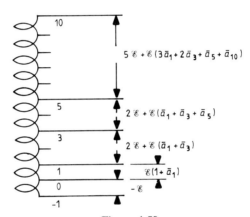

Figure 4.59

This calibration scheme has the same structure as the proposed
construction of a $+10: -1$ transformer from binary ratios described in
§4.2.3. Many other measurement schemes are possible, and a good way to
look for systematic errors in the measurement is to measure the same total
ratio using another scheme.

We now examine the various precautions necessary to realise the potential
for high accuracy of this method.

The property required of the calibration transformer T_C is that the ratio of
its secondary windings, nominally $1: -1$, shall be constant and independent
of the potential of the ratio winding with respect to the rest of the circuit. Its
electrostatic screening, and the screening associated with the detector D_1 are

critical in meeting this requirement. The ratio must also be immune to the effect of any other external perturbations, for example, vibrations or external electromagnetic fields, especially at the bridge frequency. T_C is not required to have a particularly low output impedance. A cutaway diagram of a possible design is shown in figure 4.60.

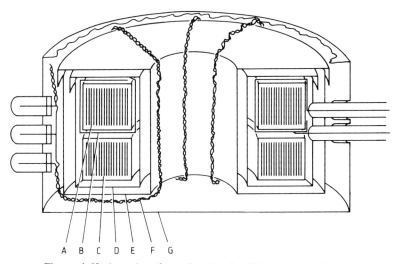

A B C D E F G

Figure 4.60 *A section through a 1: − 1 calibrating transformer.*

A is a supermumetal core wound directly with the primary winding B. The cross section of A and the number of turns of B are chosen to provide a sufficiently high input voltage rating and impedance of the transformer. C is a second core placed adjacent to A, and both are enclosed in a toroidal magnetic and electrostatic primary shield D. Two adjacent tubular peripheral limbs (shown on the right) from this shield enclose the ends of the primary winding separately. The secondary electrostatic shield E surrounds D and the twisted bi-filar 1: − 1 ratio winding F is wound on it, using sufficient turns to generate the maximum calibrating voltage needed. The ends are connected to form the three taps which are then led out in separate coaxial screens through the magnetic and electrostatic screen G which surrounds the entire structure.

By always maintaining the screens E and G at the same potential with respect to the bi-filar winding (in particular for reasons of symmetry it is desirable to maintain E and G at the potential of the centre tap of the bi-filar winding), the 1: − 1 ratio of the bi-filar windings is unaffected by being raised to various elevated potentials with respect to anything outside the space enclosed by E and G. For example, raising the potential of E and G and the enclosed bi-filar winding with respect to the primary winding B and its shield D, or to external bridge circuitry, will not affect the $1 : -1$ ratio.

4.4.3 *The bridge circuit and details of the shielding*
The bridge circuit and shielding are shown in figure 4.61. By looking at the figure it can be seen that the secondary shield of T_C is extended from T_C to enclose totally the conductors from the taps of T_C up to the points A_1, A_2 and A_3 where the output potentials of T_A are defined as those between inner and outer. A small overlap of this shield with the outers of the taps of T_A ensures complete electrostatic shielding. This shield is elevated to the potential of the centre tap of T_C by being connected to the output of IVD I_4', a very small quadrature adjustment being provided by IVD I_4, the phase-change circuit and injection transformer J_4. Establishment of the correct potential is verified by a null on the isolated detector D_3 which is connected between the central inner of T_C and the secondary shield.

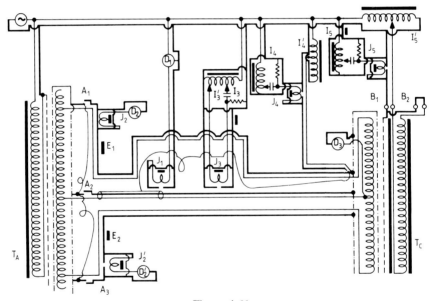

Figure 4.61

As suggested in the diagram, the main detection transformer J_1 surrounds the inner conductor from the centre tap of T_C and its shield to A_2. Since this shield is (in general) at an elevated potential, capacitive currents will flow from it to the shield of J_1 and those from the left of the split in the screen of J_1 will thread J_1 and register on D_1. By shielding the left portion with a second screen connected by a wire round the outside of J_1, the capacitive currents going from the left and right sides to the right and left sides of the split screen of J_1 can be equalised until D_1 registers a null. In practice, lack of complete equality leads to a finite detector indication, the effect of which can be eliminated by using this indication rather than a zero indication subsequently as the balanced state. Additionally, all three conductors to the right of J_1 are enclosed in a second all-embracing screen not shown in figure 4.61 so that capacitative currents which would have flowed to the individual screens are intercepted and prevented from inducing error voltages in the inners via the mutual inductance between inners and outers.

The conductors from T_C are run closely together (to avoid induced EMFs from external fields) until they separate at point A_2 to go to the taps of T_A at A_1, A_2 and A_3. An additional wire between the outers of the A_1 and A_2 sockets runs alongside the cable and both wire and cable thread the high-permeability core E_1 once; there is the same arrangement between A_3 and A_2. By this artifice, any potential existing between the outers at A_1 and A_3 is sensed and injected into the inners by E_1 and E_2 so that the circuit responds to the difference in potential between inner and outer of each of the various taps of T_A, as it is required to.

4.4.4 The balancing procedure

With reference to figure 4.61, D_3 is nulled first by adjusting I_4 and I_4'. Then D_2, D_2' are nulled by altering the overall intertap voltage of T_C to match that of T_A by adjusting I_5 and I_5'. Fine adjustment using continuously variable capacitors can help if operation of I_5 and I_5' gives undesirable surges. Lack of a simultaneous null of D_2 and D_2' is due to unequal capacitances of the cables from T_C to the defining transformers at A_1 and A_3, and can be adjusted by adding a small adjustable capacitance when measuring the $+1$, 0, -1 tap combination. D_2, D_2' are then replaced by a temporary short for rapid convergence, and D_1 is nulled by adjusting I_3 and I_3'. Successive iterated repeated balances of D_2, D_2' and D_1 rapidly lead to a simultaneous null registered by both detectors.

Because of capacitative currents, D_1 may not register a null even though zero current is flowing in the inner conductor there. The extent of this effect can be measured, in order to allow for it, by temporarily connecting at A_2 a multiport connector in which the inners are brought together at a point, as are the outers. A suitable construction is illustrated in figure 4.62, where the inners are soldered symmetrically to a disc. This provides two further ports at the potential of A_2, and the other leads of the calibration transformer are removed from A_1' and A_3 are plugged into these. Also, the primary supply to T_C is reduced to zero without altering the bridge configuration. Any resulting small deflection is nulled with I_3 and I_3' and the readings thus obtained used as corrections for this zero error to the actual readings obtained when the bridge is reconnected as drawn.

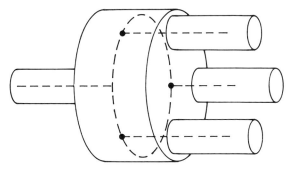

Figure 4.62

As we indicated when outlining the principle of the method, it is necessary to invert the $1:-1$ ratio and remeasure, to eliminate the small but finite departure of T_C from a perfect $1:-1$ ratio. This inversion is accomplished by interchanging leads at A_1 and A_3 and reversing the current in the primary of T_C by interchanging the input and the shorting plug at B_1 and B_2. The primary screen connection is unaltered by this reversal.

A check on the correctness of the method can be undertaken as follows. When A_2 is connected to the zero potential tap of T_A it is also possible to

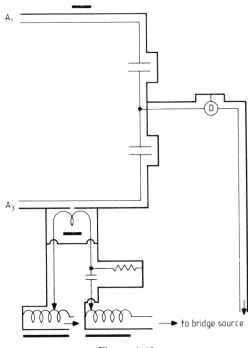

Figure 4.63

connect two nominally equal capacitors by means of T-connectors at A_1 and A_3 and use them to make a bridge. The actual ratio of the capacitors can be eliminated by reversing the connections of this capacitance bridge at A_1 and A_3, rebalancing, and taking the mean of the observations, when the departure from its nominal value of the $1: -1$ calibrating transformer should be found if the transformer bridge is balanced simultaneously with the capacitance bridge.

Then if the same value of ϵ, the departure from unity ratio of the calibrating transformer, is obtained from the transformer calibration bridge for other taps when A_2 is not necessarily at zero potential we can be confident that the calibration bridge is giving accurate results.

The circuit of a suitable capacitance bridge is shown in figure 4.63. The relative values of the capacitors must be stable over the time of observation to the accuracy required, which as we have indicated, may be 1 in 10^9 or better. A pair of 1 nF gas-filled capacitors which have nearly equal temperature coefficients and which are in the same thermostatted enclosure would be suitable.

4.4.5 Calibrating voltage transformers by permuting capacitors in a bridge

Another method of deriving ratios of voltage transformers has been described by Cutkosky and Shields (1960). This method can measure a voltage ratio nominally equal to the ratio of two integers m/n, and figure 4.64 illustrates the principle. If all the admittances are nominally equal the detector is near a null, and can be nulled exactly by the small EMF $\mathcal{E}_1$ injected in series with, say, $\mathcal{E}_n$. That is,

$$(\mathcal{E}_n + \mathcal{E}_1)(Y_1 + Y_2 + \cdots + Y_n) = \mathcal{E}_m(Y_{n+1} + Y_{n+2} + \cdots + Y_{n+m}). \quad (4.10)$$

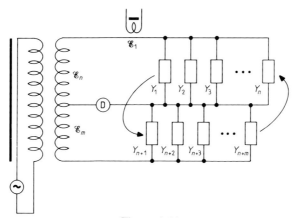

Figure 4.64

By cyclic permutation, an admittance being transferred at each step to the other side of the ratio arms as indicated by the arrows, a series of $n+m$ similar relationships to equation (4.10) can be arrived at:

$$(\mathcal{E}_n+\mathcal{E}_2)(Y_2+Y_3+\cdots+Y_{n+1})=\mathcal{E}_m(Y_{n+2}+Y_{n+3}+\cdots+Y_1)\ \text{(etc).}$$

On adding the $m+n$ relationships we have, since each individual admittance has voltage $\mathcal{E}_n$ across it n times and voltage $\mathcal{E}_m$ across it m times

$$\mathcal{E}_n n\sum_{i=1}^{n+m}Y_i+n\bar{Y}\sum_{i=1}^{n+m}\mathcal{E}_i=\mathcal{E}_m m\sum_{i=1}^{n+m}Y_i$$

where we have assumed that $\Sigma\Delta Y_j\mathcal{E}_i$ is a negligibly small sum of products. ΔY_j is the difference $Y_j-\bar{Y}$. Since

$$(n+m)\bar{Y}=\sum_{i=1}^{n+m}Y_i\qquad\text{and}\qquad(n+m)\bar{\mathcal{E}}=\sum_{i=1}^{n+m}\mathcal{E}_i$$

we find

$$\mathcal{E}_m/\mathcal{E}_n=(1+\bar{\mathcal{E}}/\mathcal{E}_n)n/m.$$

The above analysis has assumed that $\mathcal{E}_n$ and $\mathcal{E}_m$ are not greatly perturbed by the loading effects of the Y_i and their associated connecting cable shunt admittances. This is not in general true, but the complex arithmetic needed to account for such loading can be avoided by using separate current and potential leads. A suitable circuit is shown in figure 4.65 for the case of determining a $+10:-1$ ratio. The admittances used are invariably capacitors as these components have least change of value with applied

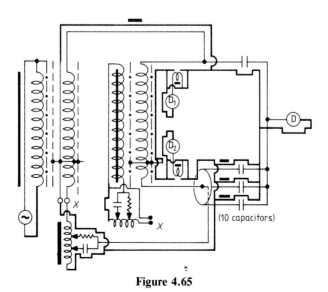

Figure 4.65

voltage. (Resistors are ruled out because of self-heating which alters their value with applied voltage on account of their finite temperature coefficients, and self-inductors are not precise standards for the reasons given in §3.5.)

Referring to figure 4.65, by means of the second core excitation and the adjustable source connected to the current winding, the detectors D_1 and D_2 associated with the defining transformers are nulled so that no current is drawn from the potential winding. From the centre of these defining transformers on through the symmetrical 'fan-out' disc and the 10 pF capacitors to the detector the admittances to the outer of each of the capacitors and associated cables are trimmed with small shunt capacitors to be equal.

The capacitors do not need to be treated as proper four terminal-pair devices with respect to the detector connections unless the differences between capacitors, which this method also yields, are required, because these cables and connectors remain unchanged throughout the permutation of the capacitors.

The voltage coefficients of the capacitors are needed to correct for the effect of the changes in voltage they experience during permutation; they have to be separately measured by the method described in §3.2.7.

4.4.6 Calibration of current transformers
In general, we will not be concerned with the calibration of these devices; a specific need will be met in §6.1.6. We can note as usual, however, that a calibration method that is suitable for a voltage ratio device can be applied to the dual current ratio device, at least in principle, by interchanging the roles of detectors and sources.

4.4.7 Assessing the effectiveness of current equalisers
These devices cannot be tested for effectiveness in themselves, but only as components of a completed bridge network. The testing philosophy is, with reference to figure 4.66, as follows.

(i) A few turns of single conductor are wound additionally around the core of the equaliser and connected to a detector. The reading of the detector is a measure of the flux in the equaliser core, which is in turn a measure of the inequality of the currents flowing in opposite directions in the inner and outer coaxial conductors.

(ii) By threading an isolated injection/detection transformer (described in §4.3.3) supplied from an adjustable current source over a cable in the same mesh of the net as the equaliser being tested, the detector can be nulled.

(iii) The current is then increased over that required to null the detector by a suitable ratio; for example one hundred times, and the effect on the main detector of the bridge network is observed. Evidently, from the linearity of

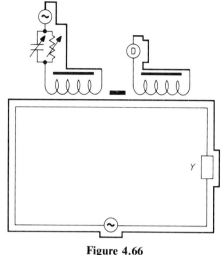

Figure 4.66

the network, the effect of the actual current imbalance in the equaliser will be 100 times less.

In this way each of the equalisers in a network can be tested in turn, and from the superposition theorem the net effect of the simultaneous imbalances in all equalisers will be the algebraic sum of the separate contributions from each equaliser.

It is an essential condition of this investigative technique that the number of equalisers in a bridge network must not exceed the minimum necessary, that is, their number should equal the total number of meshes, and that they are properly located, one to each mesh. Provided these conditions are fulfilled an equaliser may be placed in any part of a mesh. If two equalisers were to be in one mesh, then when one of them is assessed, the other would still be maintaining its function of opposing the unbalanced current and a false measure of the equaliser's effectiveness would be obtained.

A crude method of assessing whether the presence of a given equaliser is necessary to obtain the required accuracy, but which yields no quantitative information, is to wind a few low resistance turns around its core and short the ends together. If the transformed impedance of these turns is low compared to the impedance of the mesh circuit of the outer coaxial conductor, the current equal and opposite to that in the inner conductor will flow in these turns instead, and the current in the outer will be much smaller than that in the inner instead of being equal and opposite to it. Any significant change in the bridge balance brought about by the shorted turns can be observed. Remember, however, that one possible reason for no significant change being observed is the presence of another redundant equaliser in the same mesh of the network.

References

Cutkosky R D 1964 Active and passive direct reading ratio sets for the comparison of audio-frequency admittances *J. Res. NBS* **68c** 227–36

Cutkosky R D and Shields J Q 1960 The precision measurement of transformer ratios *IRE Trans. Instrum.* **9** 243–50

Deacon T A 1972 Internal admittance loading in inductive voltage dividers with 'cable' type windings *National Physical Laboratory Report DES 17*

Deacon T A and Hill J J 1968 Two-stage inductive voltage dividers. *Proc. IEE* **115** 888–92

Grohmann K and Zapf T L 1979 An international comparison of inductive voltage divider calibration methods between 10 kHz and 100 kHz *Metrologia* **15** 69–75

Hill J J and Deacon T A 1968 Theory, design and measurement of inductive voltage dividers *Proc. IEE* **115** 727–35

McGregor M C, Hersh J F, Cutkosky R D, Harris F R and Kotter F R 1958 New apparatus at the National Bureau of Standards for absolute capacitance measurement *IRE Trans. Instrum.* **7** 253–61

Thompson A M 1958 The precise measurement of small capacitances *IRE Trans. Instrum.* **7** 245–53

—— 1983 Precise calibration of ratio transformers *IEEE Trans. Instrum. Meas.* **32** 47–50

5 Sources, Detectors, Cables and Connectors

In this chapter we consider the miscellaneous, but nonetheless important properties of sources and detectors and of the coaxial cables and connectors used in bridge networks so that the practical examples of bridge design described in chapter 6 can be successfully implemented.

The chapter concludes with general precautions that should be taken, and tests which should be made, to ensure the accuracy of results obtained from bridge networks.

5.1 General principles

The waveforms which exist in the networks discussed in this book are usually closely sinusoidal and of a single angular frequency ω_0, but we can consider any periodic waveform of arbitrary shape by representing it as a Fourier series, that is, as a superposition of phase-related harmonics or multiples of the fundamental frequency whose amplitudes are such that the added result is the original waveform.

We need to consider departures from the ideal situation where

(i) the bridge source generates a pure sine wave at angular frequency ω_0,

(ii) the network is strictly linear with respect to voltage and current,

(iii) the detector responds to only a single frequency, which can be adjusted to be ω_0.

In principle, if (iii) is true, either conditions (i) and (ii) can be relaxed without incurring error, but if (iii) is not true, then both (i) and (ii) must hold if a wrong balance is not to be incurred.

5.1.1 The sensitivity of detectors to harmonic content

Bridges can be divided into two types for the purpose of analysing the departure of detectors from the ideal. Some bridges have a balance condition which is essentially independent of frequency (that is, apart from minor departures arising from the expected small frequency dependence of the components) and others have an inherently frequency-dependent balance condition. The 'twin-tee' bridge network drawn in figure 5.1 is an example of the latter, as its balance condition is $\omega RC = 1$.

If a bridge is nearly frequency independent, there will be a small amount of higher harmonics appearing at the detector terminals when the balance condition is satisfied for the fundamental ω_0. This amount will be proportional to the product of the harmonic content of the source and the residual frequency dependence of the bridge balance. If the detector is tuned to higher frequencies in the neighbourhood of the harmonics $2\omega_0$, $3\omega_0$ etc, an increase in the detector reading will indicate the presence of these higher harmonics, but the only consequence of their presence is an increase in the background indication when the detector is tuned to ω_0. The presence of this background increases the difficulty of judging the minimum indication, but this condition still corresponds to the true bridge balance. The situation can be improved by making the main balance detector more frequency selective to reject the troublesome harmonics.

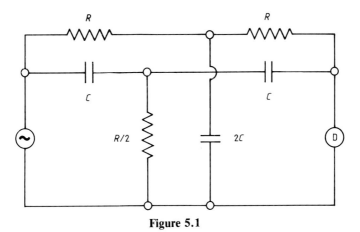

Figure 5.1

A more subtle problem arises if the amplification stages of the detector, before any frequency selection takes place, are non-linear. If, additionally, there are successive harmonics present at the input to the detector whose frequencies differ by ω_0, then the non-linear elements will produce signals at the various sum and difference frequencies, and in particular a signal actually at ω_0. An erroneous balance will result from opposing this spurious signal with an equal but opposite signal from the bridge at the fundamental frequency. Fortunately, good present-day designs of amplifiers are such that any non-linearity is rarely large enough to cause any trouble when used with essentially frequency-independent bridges.

Formally, we can see how the fundamental frequency arises from mixing harmonics if we take the detector response r to contain higher terms than linear in the applied voltage:

$$r = a_1\mathcal{E} + a_2\mathcal{E}^2 + a_3\mathcal{E}^3 + \cdots + a_n\mathcal{E}^n + \cdots \tag{5.1}$$

where usually the coefficients a_n decrease rapidly with increasing n so that only the first two terms of equation (5.1) need to be considered.

Then if the applied voltage contains harmonics

$$\mathcal{E} = \sum_{k=0}^{\infty} \alpha_k \sin[(k+1)\omega_0 t]. \tag{5.2}$$

On substituting equation (5.2) in equation (5.1), we find terms such as

$$a_2\alpha_k \sin[(k+1)\omega_0 t]\alpha_{k+1} \sin[(k+2)\omega_0 t]$$

which can be re-written as

$$a_2\alpha_k\alpha_{k+1}\{\cos[\omega_0 t] - \cos[(2k+3)\omega_0 t]\}/2.$$

Terms like the first represent the spurious signal at the fundamental frequency.

The phenomenon is called intermodulation distortion, and the magnitude of the effect is proportional to the product of the amplitudes of two successive harmonics present. A ready test for its presence is to observe whether there is any change in the bridge balance when the input signal to the detector is altered with a linear attenuator. Bridge components whose values are slightly voltage dependent will give a similar effect if the bridge source voltage is changed, and so the test of deliberately increasing the harmonic content of the source whilst maintaining its amplitude constant is also valuable in these circumstances.

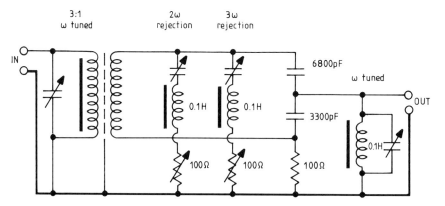

Figure 5.2 *Component valves shown are suitable for* $\omega = 10^4$ *rad s*$^{-1}$, *source impedance 100 kΩ in parallel with 1 nF. All cores are high Q ferrite material.*

When the balance condition of a bridge is frequency dependent the bridge will pass the harmonics present in the source through to the detector relatively unattenuated, whilst at balance the amount of fundamental presented should be zero. The problems of insufficient frequency selectivity and non-linearity of the detector are then far worse, and it is necessary in work of the highest accuracy to precede the detector with a frequency-selective circuit of sufficient linearity so that this circuit itself does not cause intermodulation. The magnitudes of harmonics in a source normally decrease as the harmonic order increases, so that it is usually enough to eliminate the second and third harmonics to a high degree of perfection by tuned bridge filters and to reduce higher harmonics with a simple tuned filter. Figure 5.2 shows a suitable design.

5.1.2 Noise and noise-matching a detector to a bridge network

There is an extensive literature concerned with noise (for example, see Robinson 1974). Here we discuss only some simple considerations which are often sufficient for bridge work.

It is usually easy to obtain sufficient amplification of a small out-of-balance signal from a bridge, so that the sole criterion to be investigated is the optimising of the signal-to-noise ratio of the detector. Because we are concerned in this book with bridge networks of passive components we need consider only Johnson noise ('trivial', but nonetheless troublesome noise from pick-up, microphony, etc having been dealt with). Johnson noise generates a power

$$\mathcal{E}_n^2/R = i_n^2 R = 4kT\,\delta f$$

in a resistance R. Here, k is Boltzmann's constant $\simeq 1.38 \times 10^{-23}$ J K^{-1}, T is the absolute temperature ≈ 300 K for ordinary room temperature, and δf (Hz) is the bandwidth of the detecting system, so the noise power is about 1.6×10^{-20} W for 1 Hz bandwidth at normal room temperatures. For a network connected to a detector the R concerned is the total shunt loss, that is, all sources of loss in the network as well as the detector are included in it. The performance of most modern AC detectors approaches the Johnson noise limit over a range of source impedances; optimising their performance is therefore a matter of providing appropriate matching.

The noise situation can be analysed by replacing the actual resistors and other loss sources with equivalent noise sources in series with ideal noise-free resistors (where we are considering a voltage noise source) or in parallel with them (where we are considering a current noise source). The superposition theorem then allows us to consider the effect on a detector of the signal source and the various noise sources separately.

In this and the following section, we assume that all impedances are pure resistances in order to bring out the point we wish to make as simply as possible. We also take as a simple model of a real detector the equivalent circuit of figure 5.3 where D is an ideal noise-free detector of high input impedance, that is, of negligible input admittance, and the noise and input impedances are represented by a shunt conductance G_D and zero susceptance. We suppose that the total loss of the bridge is represented by a conductance G_S, as would be the case for a bridge having totally resistive components and negligible shunt admittance.

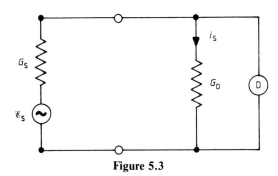

Figure 5.3

The total noise admittance seen by D is $G = G_S + G_D$. The noise power is

$$\varepsilon_n^2 G = 4kT\delta f.$$

The signal current i_S in the loop is

$$\varepsilon_S G_S G_D/(G_S + G_D) = \varepsilon_S G_S G_D/G.$$

The voltage across the detector is

$$\varepsilon_D = i_S/G_D = \varepsilon_S G_S/G.$$

Hence the signal-to-noise power ratio

$$S = \varepsilon_D^2/\varepsilon_n^2$$
$$= (\varepsilon_S^2 G_S/4)(G_S/G)(1/kT\delta f).$$

Another source of noise in addition to Johnson noise occurs at low frequencies in non-metallic conductors and semiconductor devices. Termed 'flicker noise', or '$1/f$ noise' because it is approximately inversely proportional to the frequency f for unit bandwidth, it is likely to be detectable at 1 kHz and will increase in relative importance as the frequency is reduced.

5.1.3 The concept of a noise figure

$P = \varepsilon_S^2 G_S/4$ is the maximum possible power that can be transmitted into the detector terminals, by varying G_D until it equals G_S. The signal-to-noise ratio obtained when the value of P equals the total noise power $kT\,\delta f$ is called the noise figure or factor n; it is a measure of how much the set-up fails to meet the ideal signal-to-noise ratio. In the above instance

$$n = G/G_S = (G_S + G_D)/G_S.$$

Often $N = 10\log_{10} n$ is used instead and the condition for optimum power transfer into the detector ($G_D = G_S$) gives $n = 2$ or $N = 3.01$ dB.

In the present example, it is instructive to note that

$$\left.\begin{array}{c} n \to 1 \\ N \to 0 \text{ dB} \end{array}\right\} \text{ as } \left\{\begin{array}{c} G_D \to 0 \\ \text{or } G_S \to \infty \end{array}\right.$$

that is, the best signal-to-noise ratio is *not* obtained under conditions of maximum power transfer. Thus, in general, it is important not to confuse power matching with noise matching; in a detector for bridge work it is the latter with which we are usually concerned.

For a real detector, G_D is often fixed by the instrument design; however, by interposing a matching network between the source and the detector, provided that it does not of itself introduce significant further noise, the apparent value of G_S at the detector terminals can be altered. Two such matching devices commonly used either separately or in combination are:

(i) a tuned L–C circuit (see figure 5.4(a)),

(ii) a ratio transformer (see figure 5.4(b)).

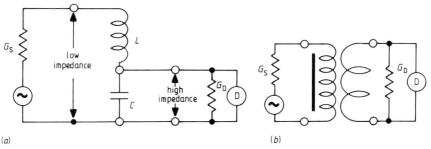

(a)   (b)

Figure 5.4

(i) can simultaneously perform the useful task of rejecting harmonic components of ε_S, whilst (ii) can provide isolation of the detector from the bridge network. Harmonic rejection can be accomplished by tuning the transformer with a capacitor which shunts either the primary or secondary winding. It is the finite inductance of each winding and the leakage inductance between them of a practical transformer which makes tuning possible, and in this respect a poor coupling between the windings can be advantageous.

From the discussion so far, it might be thought that using a matching device to make the bridge impedance presented to the detector look very low would be sufficient to resolve all signal-to-noise problems. Unfortunately this is not so because we have adopted too simple a model for a real detector; its noise attributes can rarely be represented by a simple conductance G_D.

A better representation of a real detector is that of figure 5.5 where an 'ideal' detector (of infinite input impedance and no noise) is shunted by a noiseless admittance Y_D in series with a voltage noise source ε_D, the whole being shunted by a current noise source i_D.

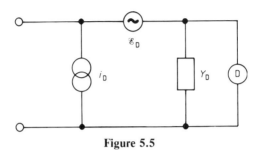

Figure 5.5

If, therefore, a detector for which this representation is valid is connected to a bridge by a matching circuit of low output impedance and low signal voltage but high signal current so that i_D and Y_D are shorted out, the detector voltage noise ε_D is still registered and swamps the low signal voltage. If the matching circuit is made to have a high output impedance with high signal voltage but low signal current, the effect of ε_D may well be negligible, but the signal current will be swamped by i_D in parallel with it. There is an optimum between these two extremes.

ε_D and i_D are easily measured for a given detector; having calibrated the sensitivity of the detector for voltage and current by connecting it to standard signal generators and attenuators, ε_D is found from the deflection observed when the input terminals are short-circuited, and i_D from the deflection observed when they are open-circuited but screened to prevent electric field pick-up.

The optimum signal-to-noise situation arises when the voltage generated by i_D flowing in the source impedance Y_S added to Y_D equals the voltage noise ε_D.

Manufacturers of detectors often supply 'noise contours', that is, loci of constant noise figures on a graph which has source frequency on one axis and source impedance on the other. From these, the optimum impedance to be presented to the detector at a given frequency can be deduced, and a suitable matching network designed.

5.2 Attributes of sources_____

The bridge source will consist of an audio-frequency signal generator, followed by a power amplifier and matching transformer if the output power of the generator is insufficient. The arrangement must be capable of delivering the necessary power without excessive distortion, but, if (as is usually the case for the bridges being discussed here) the output goes first into a transformer forming part of the bridge, there is no point in striving to achieve excessively low distortion, because, unless the output impedance of the source is very low compared with the input impedance of the transformer, the transformer will add a little distortion on its own account. A total harmonic distortion of 0.1 per cent is reasonable. The reason for the distortion added by a transformer is that the apparent instantaneous impedance of a transformer depends on the relative permeability of the core which in turn depends on the flux level, and therefore on the excitation level. Variation of the instantaneous input impedance causes varying voltage drop in the output impedance of the source, and hence distorts the waveform.

If the bridge balance is frequency dependent the source needs to have the necessary frequency stability; this is not difficult if the source is a frequency synthesiser with a crystal oscillator controlling the output frequency, or is an oscillator capable of being phase-locked to a synthesised frequency. The frequency can be measured by a suitable electronic counter operating in the period mode.

5.3 Properties of different detectors _____

5.3.1 Preamplifiers and matching transformers
As already discussed, these items enable the bridge network to be noise-matched to the detector; in general, negligible additional noise is introduced by well designed matching transformers at impedance levels below $10^6 \, \Omega$, provided that they are constructed with high-permeability low-loss cores. The output of a matching transformer may need further amplification and transistors having noise figures below 4 dB can be used at the lower impedance levels concerned. At higher impedance levels electronic amplifiers using low-noise field effect transistors are effective.

A particularly difficult matching and noise problem arises in connection with transformer bridges designed to measure the ratio of small-value capacitors. The application to a calculable capacitor is an extreme case where a direct capacitance of less than 1 pF at $10^4 \, \text{rad s}^{-1}$ must be observed with a resolution approaching 1 in 10^9 in the presence of up to 1 nF of shunt capacitance. At a working frequency of $10^4 \, \text{rad s}^{-1}$ these impedances are of the order of $10^8 \, \Omega$ and $10^5 \, \Omega$ respectively. At least three different solutions have been successfully adopted in different laboratories. Cutkosky (1968) at the National Bureau of Standards, USA, used a semiconductor parametric amplifier which he cooled to 77 K to reduce the thermal noise. Ricketts (1978) at the National Measurement Laboratory, Australia, developed a superconducting, very low-loss inductor which was tuned by the shunt capacitance. The combination presented the necessary high impedance to the capacitor, and the inductor was tapped to provide a low-impedance output to a conventional amplifier. This solution is very attractive, but it does depend on the ready availability of liquid helium.

We have found that the amplifier designed by Harkness (1977), shown in figure 5.6, which is based on a low-noise field effect transistor, yields very good results, particularly when one special type of ultra-low-noise field effect transistor, BF817, is used. Other transistors in the series, BF800, 805 and 806 are useful for matching to smaller input capacitances. The noise performance is comparable with that of the above two solutions and the device is a very convenient head amplifier for high impedance bridges in general.

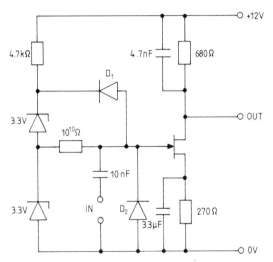

Figure 5.6 *A low noise preamplifier suitable for capacitance comparison bridges.*

The diagram shows just the input stage, including overload protection diodes, which should have a leakage current of less than 2 pA. These enable overload conditions encountered, for example, when one component is removed from a bridge without reducing the source EMF to zero, to be withstood without harming the amplifier. To achieve the lowest possible noise the inevitable stray capacitances around the input should only include air or low-loss dielectric material such as PTFE. The voltage gain of this input stage is not great; the low noise properties result from impedance matching as the input would typically be connected across a low-loss 1 nF shunt capacitance. The output is suitable for following by a conventional low-noise transistor amplifier. A somewhat similar solution, using a 2N 6550 FET transistor, has been described by Ramm and Bachmair (1982), and the complete preamplifier is commercially available.

5.3.2 Wideband (untuned) detectors
The most useful example of this type of detector is the oscilloscope, whose use may well reveal the presence of unwanted interfering signals on the output. There have been various schemes which used an oscilloscope whose x-deflection is driven directly from the bridge source as a type of phase-sensitive detector. Apart from revealing the direction of bridge unbalance when rather far from balance without saturating, because the display is then a line inclined a little one way or the other to the vertical, this method would seem to have no advantage over a conventional phase-sensitive detector. The necessary narrow bandwidth for a good signal-to-noise ratio has to be obtained by preceding the oscilloscope with a narrowband tuned pre-amplifier.

5.3.3 Narrowband (tuned) detectors
These often incorporate a frequency-selective network (usually a twin-tee bridge) in the feedback loop of an electronic amplifier. The frequency response is generally similar to that of an $L-C$ tuned circuit, that is, it rolls off away from the resonant frequency at a rate governed by the effective Q of the circuit. Successive harmonics are attenuated with respect to the preceding one by a factor of the order of Q. It is difficult to design stable circuits with Q values much greater than 10^2-10^3, so the detector will have for the second harmonic $2\omega_0$ not much less than a part in a hundred of the sensitivity which it has for the fundamental and as we have seen, this situation can cause errors in setting up the bridge balance condition.

Further, a simple tuned detector responds to both in-phase and quadrature signals alike; the deflection corresponds to the root-sum-of-squares of the in-phase and quadrature components. This can make the balancing of a bridge where the in-phase and quadrature balances are not independent somewhat difficult. A degree of phase dependence can be obtained by deliberately introducing, directly into the detector from the bridge source, a signal of the desired phase. Smaller signals in quadrature with this will then not produce much deflection, but in-phase signals will be fully displayed as an addition to the standing deflection.

There is a useful technique for improving considerably, by about a factor of ten typically, the sensitivity of such detectors to a signal which approaches the detector noise in magnitude when the signal comes from a nearly balanced bridge. It is to take as the balance point the mean of two adjustments on either side of the bridge balance condition which produce equal, relatively large, deflections of the instrument rather than to try to find the bottom of a shallow minimum in the detector deflection. The response to a signal decreasing to zero becomes parabolic because of

(i) the presence of background detector noise,

(ii) the detector square-law response for small signals,

(iii) as we have noted above, the presence of any residual component in quadrature with the adjustment being made.

By using this technique, one of the advantages of phase-sensitive detection can be obtained with a simple tuned detector, namely that the balance condition for the adjustment of one phase component is not obscured by a maladjustment of the bridge for the signal in quadrature.

5.3.4 Phase-sensitive detectors employing a switching technique

All phase-sensitive detectors use a reference signal; in bridge work it is usually derived directly from the bridge source. Many phase-sensitive detectors apply the reference signal to switching circuitry which rectifies the input signal, others dealt with in the next section use it to modulate the input signal. In the rectifying type, since any harmonics present in the input signal are phase-related, they also will be rectified. For even harmonics there is no net rectification in the first half-cycle of the reference waveform, and so these harmonics are not detected, but the last one of the n cycles of the nth odd harmonic does not cancel, so that the odd nth harmonic is detected with $1/n$ efficiency relative to the fundamental. Therefore phase-sensitive detectors which use a switching principle need to be used with caution on bridges whose balance condition is nominally frequency independent and must be preceded with a tuned filter when used with a frequency-dependent bridge.

5.3.5 Phase-sensitive detectors employing a modulating technique

Other types of phase-sensitive detectors in effect multiply the incoming signal with the reference signal and display the result after passage through a low pass filter. The effect for all frequencies applied to the signal input other than that of the reference signal is a time-varying output at the difference frequency which is rejected by the low-pass filter. Therefore there is, in principle, complete rejection of signals at all frequencies except for the one desired. Even the harmonics of the desired signal are rejected. It might therefore be supposed that this kind of phase-sensitive detector is the most suitable for bridge work. Unfortunately, particular examples may be prone to non-linearity at small signal levels and may also have limited amplitudes of acceptable total input signal plus noise before their saturation level is reached. Both defects lead to a greater likelihood of intermodulation distortion.

5.4 Cables and connectors_____

The properties of coaxial cable used for high-accuracy work need careful consideration. Where bridges for comparing small direct admittances (such as small-value capacitors) are concerned, any loss in the cable dielectric (particularly of cables making connections to the detector or associated network) will add noise. The presently available most useful materials for the dielectric between inner and outer conductors are polythene and PTFE. If the capacitance per unit length of the cable is unduly large, the available out-of-balance signal will be shunted away, and the balance sensitivity will be decreased. Unfortunately, cables with lower values of shunt capacitance inevitably have high series inductance, as the two are linked by the characteristic impedance of the cable, and high inductance and the accompanying high series resistance means large corrections to the two and four terminal-pair values, as was pointed out in §2.4. These effects are exacerbated in those types of cable where the central conductor is of copperweld construction to increase its mechanical strength. A copperweld conductor has a steel core sheathed by a thin layer of copper.

All coaxial cables used at audio frequencies suffer from microphony; in an extreme case, the acoustic oscillations of a cable carrying a high-level signal can be transmitted to a touching cable associated with the detector, and can then be reconverted into an electrical signal. Therefore all cables should be of non-microphonic construction; this is usually accomplished by putting a layer of conducting plastic between the dielectric and the outer conductor, as friction between these is the principle cause of microphony. Such a layer also prevents leakage of electric field through the holes in the woven outer conductor. The resistivity of the plastic layer can be unduly high in some cables and this can give rise to an additional loss mechanism and consequent Johnson noise in the cable. This type of cable is not suitable for high frequency applications, as the conducting layer gives a high loss.

The choice of cable for use in accurate bridge work at audio frequencies involves a compromise between low shunt capacitance, low series inductance, low series resistance, reasonable physical size and flexibility, and availability. The authors have made extensive use of a PTFE insulated cable of about 6 mm overall diameter, having a characteristic impedance of 75 Ω, shunt capacitance of 70 pF m^{-1} with an associated loss angle of 35×10^{-6} rad and a series inductance and resistance of 0.4 μH m^{-1} and 0.064 Ω m^{-1} respectively. Microphonic effects are reduced by a coating of conducting varnish applied to the dielectric. This varnish is easily removed locally when terminating the cable with a connector.

The principal requirements for coaxial connectors are that the series resistance they introduce between their mating surfaces should be low, as should be the loss in any dielectric used in their construction, the inner conductor should be completely screened when the connectors are mated, and there should be available T-connectors of low impedance at audio frequencies. Laboratories undertaking work of the highest accuracy all use the British Post Office MUSA connectors, which, besides the above desirable properties, are also simple to attach to cable ends and components, are relatively inexpensive, and mate and unmate with a simple push and pull action. The more readily available BNC type of connector suffers from a higher and more variable contact resistance, but might be acceptable for lower accuracy work, or where series resistance is in quadrature with the principal impedance of interest, that is, a capacitance or inductance, or in high impedance networks at moderate accuracies. Ordinarily available cables with ready-fitted BNC connectors do not have a conducting layer and are not suitable where low noise is important. GR900 connectors are unnecessarily elaborate and therefore expensive for audiofrequency applications. The hermaphrodite GR874 type are suitable, but care must be taken to ensure that the central mating conductors do not become partially unscrewed and provide a source of unwelcome variable resistance. The screening of the inner by the outer is also not quite complete, but this is not usually of any significance.

5.4.1 An adaptor to convert two and four terminal defined components to a two or four terminal-pair definition

There are valuable standards still in use which are provided with only two or four single terminals of the exposed screw binding-post type, plus possibly a screen connection. If the admittance of such a standard is large, the uncertainty caused by the electric field in the vicinity of these terminals may be acceptably small, but care has to be taken in accounting for the magnetic fields caused by the currents flowing in the connections to the standard, and we still need to take precautions against the effect these fields might have on the rest of the bridge network. These problems may be overcome by constructing the rest of the bridge network in a proper coaxial manner, and by making an adaptor to convert these kinds of standards to a two or four terminal-pair definition which leads to a value not significantly different from their two or four terminal defined value.

We describe first an adaptor for use between a four terminal-pair bridge and a four terminal component, and then describe how this same adaptor can be used to connect two or four terminal-pair bridges to two or four terminal components. The adaptor is shown in figure 5.7.

From the point of view of the four terminal-pair bridge network, the two problems are firstly to bring the outers of the four cables directed at the component together in such a way that their junction possesses zero four terminal impedance, and secondly to continue their inners to the terminals of the device in such a way that there is acceptably small mutual inductance and capacitance between these leads.

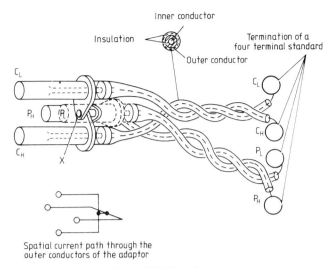

Figure 5.7 *The adaptor*.

The outers can be dealt with by joining them using a four terminal zero-impedance junction. A zero impedance has the property that the potential between its potential terminals is zero when a current flows between its current terminals (this is also a description of a balanced bridge network). A zero resistance for DC use need merely have a conductor joining its current terminals and another joining its potential terminals with a conductor of small cross section joining points on each of them. For AC use it is also necessary that there is no significant mutual inductance between the current and potential conductors, and this can be done by a suitable geometrical arrangement of the device.

The inners of the adaptor need to be continued as separately screened, insulated wires twisted together in current and potential pairs for a sufficient length to give flexibility so that they can be separated and led to the separate terminals on the component in spatially orthogonal routes. An implementation of these principles using British Post Office MUSA connectors is sketched in figure 5.7. The coaxial connector P_L nearest to the reader as well as the inner connectors have been outlined in broken lines to enable details of the construction to be seen. The outers of P_H and P_L are soldered together at X. A conducting bolt goes through an insulating spacer to the outers of C_H and C_L which are similarly joined; in one mode of use a non-conducting bolt isolates the P outers from the C outers. The geometrical configuration of the current path through the outers has the symmetry of the inset to figure 5.7.

The adaptor is shown connected to the terminals of a four terminal component whose terminals are simple binding-posts. When connected to a two terminal component, the leads C_L and P_L should be connected to one terminal and C_H and P_H to the other. For conversion to a four terminal-pair all four coaxial connectors are used (see table 5.1). If the component has a screen to be maintained at a low potential, it should be connected to any outer with a separate wire.

Table 5.1

		To convert to	
		Two terminal (§2.2.3)	Four terminal (§2.2.4)
To convert from	Two terminal-pair (§2.2.6)	Use C sockets and leads only	Isolate C outer from P outer, short C_L, P_L sockets
	Four terminal-pair (§2.2.8)	Connect C_L, P_L leads to low terminal of standard, C_H, P_H to high terminal	Use adaptor as illustrated (figure 5.7)

5.5 General points concerning the accuracy of bridges

5.5.1 *Avoiding unwanted interactions between measuring circuits*

This topic is of the utmost importance, not only with regard to the subject matter of this book, but also throughout the whole field of electrical measurement where results need to be free from inaccuracy and signal-to-noise problems caused by power line pick-up, interference from other external sources and unwanted interactions between different parts of the system. A 'system' could be a network of impedances and instruments, or the internal workings of a single instrument.

Of the plethora of books and articles on this subject, some are helpful whilst others confuse the issue. Unfortunately, skill in avoiding troubles of this kind is often regarded as a black art, although the principles involved are straightforward and simple. In this section we outline these principles, and in the following sub-sections we discuss a particular problem which arises in bridge networks.

In §1.4 we emphasised the importance of examining the currents which flow round *complete* circuits. These currents will give rise to electric and magnetic fields which will afford further opportunity of interaction. We can classify these modes of interaction as follows.

(1) Common impedance. The two circuits share a common section which has a finite impedance (figure 5.8). The potential drop down this caused by the current i flowing around *one* circuit may be regarded as a source in its own right which can drive unwanted current i' around the other circuit.

This situation can be cured either by reducing the common impedance to zero by using a star connection, as in figure 5.9, or a branch connection, as in

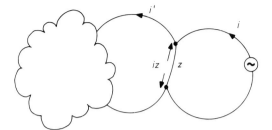

Figure 5.8 *Circuits coupled by a common current path.*

figure 5.10, to join the two circuits. Another possibility is to use the concept of a combining network discussed in §2.7.

(2) Either two capacitances, or more commonly, one capacitance and a linking conductor (which is frequently provided by the green–yellow or 'earth' wire of the mains supplies) can couple two systems via their external electric fields, as in figure 5.11(*a*) or (*b*). Many instruments and bridge components may appear to be encased in fairly complete conducting enclosures but this can be illusory if the panels of these enclosures are in fact insulated from one another by, for example, an oxide layer or paint. Also, the shafts which couple internal controls such as switches and potentio-meters to knobs outside the case are not always connected to the case and

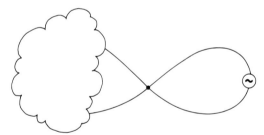

Figure 5.9

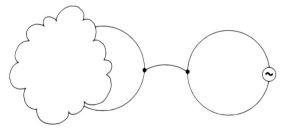

Figure 5.10

therefore provide a means of allowing currents via electric fields to enter or escape from the case.

The cure for coupling via an electric field (capacitative coupling) is either to intercept the electric field with a conducting shield connected into the system in such a way as to return the current harmlessly back to its source, or to arrange that the parts liable to be coupled in this way are at the same potential.

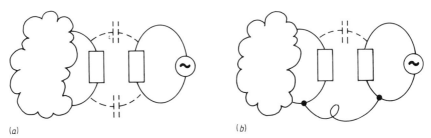

(a)   (b)

Figure 5.11 *Circuits coupled by capacitance.*

(3) Complete meshes of networks can be coupled via the magnetic field arising from current flow around one of them (figure 5.12). That is, circuits can be coupled by mutual inductance. This coupling can be reduced by decreasing the effective area of one, or preferably both, of the meshes, for example, by using twisted wire conductor-pairs, or the coaxial conductor-pairs which are the subject of this book. If there are multiply connected conductor-pair meshes, then they will need to be current equalised.

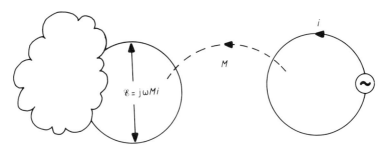

Figure 5.12 *Circuits coupled by mutual inductance.*

Finally, multiple causes of interference which have a primary common source so that they are phase-related, being all at the same frequency, add up vectorially and the final resultant may well be, accidentally, quite small. In that case, elimination of only some of the coupling modes may well appear to make the resultant bigger. It is most important to realise that this apparently worsened situation is illusory, and to go on and remove the remainder of the couplings.

5.5.2 Single conductors added to an equalised network

In this section we outline a general principle and in the following section we give an important application of it to bridge work.

The currents flowing in the network of low-impedance coaxial outer conductors of an equalised conductor-pair net cause small but finite potential differences between parts of these conductors. The currents flowing because of these potential differences through the small capacitances between the different parts of the outer conductors are usually too small to cause any difficulty, but one exception that sometimes occurs arises from the relatively large interscreen capacitance of an isolating transformer when the balance of an associated bridge network is detected by a mains energised detector joined by a common single conductor (the green–yellow wire of the mains supply) to a mains energised source. This is an example of case (2) of the preceding section.

Now, in general, two points of the outer conductors of a coaxial equalised conductor-pair network can be made to have the same potential if we add a single, low-impedance conductor between these points together with an equaliser in the mesh thereby added to the original network. The magnetising current of the equaliser flows in the extra conductor, and the equalisation of the currents in the original mesh is improved. A single unpaired conductor of this kind may be brought into the philosophy of equalised conductor-pair networks by regarding it as the low-impedance member of an equalised pair whose (non-existent) partner has infinite impedance.

In figure 5.13 the points a and b have been joined by a single conductor c, and the equaliser added to the original mesh.

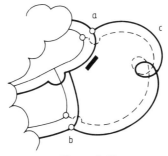

Figure 5.13

5.5.3 Meshes connecting sources and detectors which are additional to a bridge network

A particular example of the general problem treated in the last section is the connection of a mains-energised detector to detect the balance of a bridge network, when the primary bridge source is also mains energised. The problem is exacerbated when the detector is of the phase-sensitive type, and therefore needs a reference signal, which is usually taken directly from the source. The relevant parts of the network, including the green–yellow wire of the mains which also connects source to detector, are shown in figure 5.14. There are two undesired currents. Firstly, if the two screens of the transformer are not quite at the same potential, a current will flow across the capacitance between them, through the network of outer conductors in the bridge, through the outer of the cable to the detector input and returning to the transformer primary and its screen via the green–yellow wire and the reference supply cable outer. Secondly, a proportion of the reference signal current will return to the source via the green–yellow wire and a small proportion via the detector input lead outer conductor, the bridge network outers and the interscreen capacitance of the transformer. The currents which flow from both of these causes down the outer conductor, but not the inner, of the input lead to the detector cause a potential drop down it and so the condition $\varepsilon = 0$ between inner and outer conductor which holds at the detector input when it registers a null does not hold where this input cable joins the bridge network, and this leads to an incorrect balance condition being established.

The net current in the detector lead can be detected by the total current detection transformer described in §4.3.3, and may be reduced to an

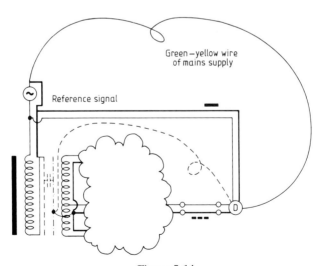

Figure 5.14

acceptably low value in either of two ways. An additional single conductor may be used to join the secondary screen of the transformer to the outer of the detector case, and an equaliser put in the cable to the detector input. Both these additions are shown as broken lines in the diagram. Alternatively, an auxiliary small voltage source comprising an injection transformer and associated circuit (§4.3.4) may be connected between the primary supply outer and the primary screen to modify the potential of the latter until it is equal to that of the secondary screen, so that no current flows between them. The achieving of this condition may be detected by the observation of no net current in the cable to the detector with a total current detection transformer.

5.5.4 Tests of the accuracy of bridges

Given good quality standards, it is relatively easy to assemble and balance a bridge, which, if proper attention has been paid to consideration of the signal-to-noise ratio, has a very high resolution. One can discern changes as small as 1 in 10^9, with a similar degree of stability of the balance. In this situation it is rather easy to be misled into believing that the results obtained also have a similar accuracy. There are a number of general tests which should be applied to any untried network to uncover possible causes of systematic error. For each test, the bridge is supposed to be fully balanced in its normal state with the source EMF and detector sensitivity at the highest level at which the bridge network will be used, unless otherwise stated. We assume that the important injected voltages have been calibrated (§4.3.6), and the equalisers tested (§4.4.7).

Test	Desired result of test	Possible reasons why desired result is not obtained
(1) Main bridge source removed, replaced by a short.	Main detector reads zero (Johnson noise excepted)	Pick-up of electric or magnetic fields or net current flow (inadequate isolation by transformer screens) or acoustic coupling.
(2) Detector gain decreased from its highest setting.	No change of zero detector reading	Same causes as for (1). Intermodulation distortion.
(3) Source output varied over at least 2:1, keeping the relative harmonic content constant.	No change of zero detector reading	Transformer ratio or impedance with a voltage-dependent value. (Demagnetise?) Intermodulation distortion.

Test	Desired result of test	Possible reasons why desired result is not obtained
(4) Source output constant, harmonic content varied.	No change of zero detector reading	Frequency-dependent bridge, inadequate frequency selectivity of detector. Intermodulation distortion.
(5) If a simple ratio bridge, measure two zero admittances.	Zero detector reading	Inadequate number or placement of equalisers. Auxiliary balances insufficiently well understood.
(6) Measure two zero impedances. (NB source matching may need altering.)	Zero detector reading	As for (5).
(7) Shorted turns around core of each equaliser in turn.	A definite detector deflection	(i) currents in inner and outer conductors by chance balance in that mesh—retain the equaliser. (ii) more than one equaliser in the mesh. Remove all but one.
(8) Open coil fed from bridge source passed near bridge cables and components. (NB a sensitive test; assess what is a tolerably small response. Very useful for low impedance circuits.)	Zero detector reading	Inadequate magnetic screening of components. Meshes without equalisers.
(9) Metal plate fed with high potential from source, passed near bridge cables and component.	Zero detector reading	Cable outers or component cases are inadequate electrical screens. Unused coaxial ports without a conducting cover.

A two terminal-pair component which has zero admittance is constructed as illustrated in figure 5.15 where a break in the inner conductor of a coaxial cable has a complete screen across the break to intercept electric fields.

Figure 5.15

A two terminal-pair component which has zero impedance requires no potential to be generated across one port no matter how much current flows in and out of the other. It is important not to include any mutual inductive coupling in the direct impedance. Figure 5.16 illustrates a suitable arrangement.

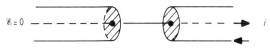

Figure 5.16

Four terminal-pair versions of the two terminal-pair components merely need the addition of two extra ports.

References

Cutkosky R D 1968 A varactor null detector for audio frequency capacitance bridges *IEEE Trans. Instrum. Meas.* **17** 232–8

Harkness S 1977 Low noise FET amplifiers for low loss capacitative sources *National Physical Laboratory Report DES 41*

Ramm G and Bachmair H 1982 Optimisation of the signal-to-noise ratio of AC bridges *Tech. Mess.* **49** 321–4

Ricketts B W 1978 Capacitance bridge null detector with a superconducting inductor *J. Phys. E: Sci. Instrum.* **11** 635–8

Robinson F N H 1974 *Noise and fluctuations in electronic devices and circuits* (Oxford: Clarendon)

6 Examples of Bridges for Impedance Comparison

We now bring together all the concepts and devices described so far to show how they may be combined to make practical high-accuracy bridges. The number of networks which may be devised is endless and therefore we have chosen just those examples which are either of considerable practical importance or illustrate the use of some device or concept. In principle we show how to relate like impedances of different value and then how to generate the impedance elements of R, L and M from the fundamental unit of capacitance. The reader should feel encouraged to devise other networks to serve other special needs.

It is certainly a major feature of the remarkable elegance of the subject that, in principle, bridge comparison networks can be built from just five dual pairs of components. These we have introduced in an *ad hoc* manner, so that the significance of the pattern may well not have been apparent. The pairs are:

(1) Impedances/admittances
(2) Sources/detectors
(3) Voltage/current ratio transformers
(4) Injection/detection transformers
(5) Combining networks/current sources.

We will often present a network as an admittance bridge; of course the distinction between this and an impedance bridge is purely one of algebraic convenience.

Although it is possible to construct any network, and in particular any of the classical four-arm bridges in conductor-pair form, we do not give many examples of this; instead it seems more valuable to make use of the ready availability or ease of construction of transformers of variable or fixed ratio to look at measurement problems afresh and, starting from the defining conditions for the components to be compared, design appropriate networks which make use of the desirable properties of transformers.

Variable components (for example, decade resistance boxes, variable capacitors and switched tapped mutual inductors) were formerly used

170

exclusively for adjustable components. Whilst these are still useful, especially since many laboratories still have a good stock of them, we in general advocate the use of a good quality admittance of fixed value supplied from an adjustable voltage source (an IVD) as the combination has a greater stability against time and environmental effects.

In bridge networks which need a detector in more than one position, only one actual detector need be used, connected into the various detector positions in turn to make the associated adjustments.

6.1 Bridges to measure the ratio of like impedances_____

6.1.1 A two terminal IVD bridge

In figure 6.1, $A_{1,2}(1 + jD_{1,2})$ are the admittances to be compared. When the detector current is zero, we can equate currents into the node to which the right-hand side of the detector is connected:

$$(\varepsilon - n\varepsilon)A_1(1 + jD_1) + jA(k - n)\varepsilon = n\varepsilon A_2(1 + jD_2).$$

Equating real and imaginary parts

$$A_1 = A_2 n/(1 - n) \tag{6.1}$$

and

$$D_1 = D_2 + (A/A_2)(n - k)/n. \tag{6.2}$$

Equations (6.1) and (6.2) express an unknown A_1 in terms of the known standards A_2 and A.

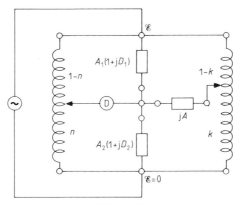

Figure 6.1

For example, if components 1 and 2 are slightly lossy capacitors with small loss tangents D_1 and D_2

$$A_{1,2}(1 + jD_{1,2}) = j\omega C_{1,2}(1 + jD_{1,2}).$$

jA would then be a conductance $G = 1/R$.

The bridge equations relate the capacitance values

$$C_1 = C_2 n/(1 - n)$$

and the loss tangents

$$D_1 = D_2 + (\omega C_2 R)^{-1}(k - n)/n.$$

The last equation represents a frequency-dependent quadrature balance, but as standard capacitors usually have very small loss tangents this does not cause any difficulty because this balance is much less critical than that relating the capacitance values.

Note the connection of the source and the IVDS to the defining terminals (§2.2.6) of the two terminal components. The current through the components is drawn directly from the source, and the IVDS, being high input impedance devices, draw little current. Hence there is only a small potential drop along the connecting wires between the IVDS and these terminals.

6.1.2 Two terminal-pair IVD bridge

This bridge, shown in figure 6.2, is suitable for comparing the ratio of two like admittances. It is derived from the bridge described in the previous section by using the coaxial techniques developed previously.

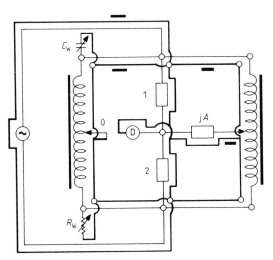

Figure 6.2 *A two terminal-pair bridge based on IVDs.*

Notice that the shorted variable output of the left-hand IVD is now labelled as the reference point 0 for convenience. The small potential drops incurred by the currents returning along the outers to 0 from the point where the isolated detector is connected become part of the measurement; each mesh of the network has been provided with a current equaliser to ensure the good definition of these currents and potentials.

The shunt admittance of the cables from the upper and lower output ports of the left-hand IVD to the components 1 and 2 are loads on these outputs which affect the output voltages of an IVD having finite internal winding resistance and leakage inductance. Wagner components R_W and C_W overcome this problem by balancing these admittances. C_W and R_W, either of which may need to be connected in either position, are to be set by putting the detector in the port 0 (the short having been removed) and adjusting C_W and R_W to attain a null.

When accurate calibrated IVDs are used in this simple bridge it is capable of a measurement uncertainty of less than 1 part per million for moderate values of $|Y_2|/|Y_1|$, say $10:1 > |Y_2|/|Y_1| > 1:10$.

6.1.3 A four terminal-pair IVD bridge

We now describe how, with but little extra elaboration, the network of the previous section can be adapted to compare two like impedances defined in a four terminal-pair manner. This network is especially useful for comparing low impedances for which four terminal-pair definition is necessary to obtain the required accuracy. By using the adaptor described in §5.4.1, two and four terminal components can also be measured on this bridge.

The circuit is shown in figure 6.3 except that, for simplicity, we have omitted to show any Wagner components. Often, in view of the other approximations to be made, their use will not be necessary. The theory of the Kelvin double bridge given in §2.7.2 is applicable, and we make use of the combining network described in §2.7.3 to connect in series the two components to be compared.

In this network we also introduce a method of making the quadrature balance by injecting into the main IVD output a quadrature voltage derived from a second IVD and phase-changing network of the type discussed in §4.3.4.

The bridge is balanced by first making the value of z small and adjusting the main and quadrature IVDs in turn to null the detector. Then z is increased to be of the same order of magnitude as Z_1 and Z_2 and the detector is again nulled by adjusting the settings of the combining network IVDs. This procedure is iterated until a null is obtained whether z is inserted or not. The null obtained with z small is the critical one, and the adequacy of the auxiliary balance of the combining network can be tested with z small by changing the settings of the combining network by an amount comparable with, but larger than, the fineness of their adjustment and observing whether

combining network   main ivd   quadrature ivd

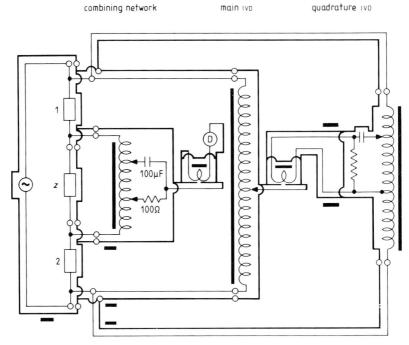

Figure 6.3 *An IVD-based bridge for comparing four terminal-pair components.*

the main balance is significantly affected. The principle of this testing procedure is very important and should always be invoked to test the satisfactory adjustment of the auxiliary balances of a bridge.

The chief approximations involved in this network arise from the voltage drops along the cables to the main IVD. The principal voltage drop arises from the magnetising current of the single stage IVD flowing through the resistance and inductance of these cables; this error source which causes large quadrature errors at low frequencies can be largely overcome by using a two-stage IVD (§4.3.2) with its magnetising winding driven directly from the source. There remains the inevitable voltage drop caused by the current flowing through the shunt capacitance of the cables and the series impedance $R/2 + j\omega L/2$ of the cables (§2.4.4), in addition to which the difference of the currents flowing through the shunt admittances of these cables causes voltage drops in the output resistance and leakage inductance of the IVD. The cables to the combining network have a very small potential difference between inner and outer, and do not contribute significant error compared to the above causes.

These errors limit the accuracy of the network to perhaps about 1 in 10^7, but the network is useful for comparing low-impedance standards. The resolution obtainable is adequate as the detector transformer also provides impedance matching between a low bridge impedance (if large admittances are being compared) and the comparatively high input impedance of a typical detector. The quadrature balance is provided by injecting a quadrature voltage $j\beta\varepsilon$ in series with the IVD output, and the bridge balance equations follow straightforwardly from equating the division of potential ratios of the components and the IVD together with the injected voltage. They are, in the notation of §6.1.1, if $D_{1,2} \ll 1$,

$$A_1 = A_2 n/(1-n)$$

and

$$D_1 = D_2 - \beta/[n(1-n)].$$

6.1.4 A two terminal-pair bridge using a 10: −1 voltage ratio transformer

The comparison of two like admittances which have a ratio of values very close to 10:1 is a frequently encountered problem. A 10: − 1 voltage ratio transformer (§§4.3.4–5) which has been calibrated (§§4.4.1–5) may be used as shown in figure 6.4, and the small additional variable voltage needed to

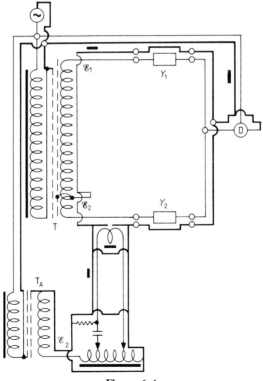

Figure 6.4

balance the bridge can be obtained by using an injection transformer, driven from an isolating transformer (§4.3.6) to preserve the isolation of the bridge by the two electrical screens of the voltage ratio transformer.

The transformer ratio (nominally 10: − 1) is defined as the ratio $\varepsilon_1/\varepsilon_2$ where ε_1 and ε_2 are the voltages appearing across the upper and lower exit ports of the transformer under conditions of zero current flow through these ports. For moderate admittances $Y_{1,2}$ and cable shunt admittances this ratio will not be greatly affected by the finite current drawn from low output impedances. If the inaccuracy this introduces is unacceptable, the loading effects can be corrected for, or the four terminal-pair values of $Y_{1,2}$ can be obtained by an extrapolation technique (§2.3) at the expense of performing one or more extra balances and doing the extrapolation arithmetic.

The potential presented to Y_2 is $\varepsilon_2(1 + \alpha + j\beta)$ where $\alpha\varepsilon_2$ and $\beta\varepsilon_2$ are the in-phase and quadrature injected voltages, so that the equation

$$\varepsilon_2(1 + \alpha + j\beta) Y_2 = \varepsilon_1 Y_1$$

holds when the detector is nulled.

Notice that if T and T_A are provided with two screens connected as shown, the measurement network is well isolated from the source and primary circuit (§4.3.6) permitting the use of a phase-sensitive detector which requires a direct connection to the source for its reference input (§§5.3.4–5, 5.5.3).

6.1.5 A four terminal-pair bridge which uses a two-stage 10: − 1 voltage ratio transformer

We now describe a network in which both the defining conditions for the transformer output potentials and the four terminal-pair definition of the components to be measured are fulfilled. An increased complexity is involved, but nevertheless achieving the multiple detector balances is straightforward.

The principle of the bridge is shown in figure 6.5. The like impedances Z_1 and Z_2 are to be compared: their nominal values are such that $Z_1 \simeq 10Z_2$. They are defined by the four terminal-pair convention of §2.2.8, although for clarity in this diagram the outer conductors of the network are not shown.

The all-important ratio of EMFs across the impedances is measured under conditions of zero current at the defining points of the leads. It is varied to achieve balance by injecting a small, known voltage $\Delta\varepsilon$. In order to achieve simultaneously zero current at the output of the two potential windings, two adjustments are necessary. The overall EMF across these windings is altered without altering their ratio by altering the flux threading them both. This is achieved by injecting a current into a winding on a second core, which is not shown. The independent other adjustment is provided by the adjustable current source.

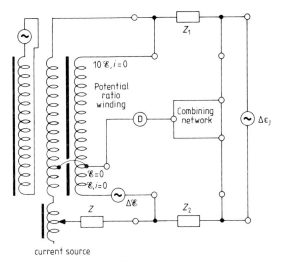

Figure 6.5

At the low potential end of each impedance, the defining current leads are joined with a means of injecting a voltage $\Delta \varepsilon_J$. The bridge is made insensitive to the inevitable voltage drop down these current leads by removing the main bridge source, injecting $\Delta \varepsilon_J$, and balancing the combining network so that the detector does not respond to $\Delta \varepsilon_J$. After making this auxiliary balance, $\Delta \varepsilon_J$ is set to zero, and if necessary $\Delta \varepsilon$ readjusted. Convergence is very rapid, however, and few iterations of these balances are necessary to make them simultaneous.

A full conductor-pair diagram is drawn in figure 6.6. The component parts can be readily identified by comparison with the previous diagram. The detectors $D_{1,2}$ and associated detection transformers when nulled ensure that the condition of zero current at the outputs of the potential windings is satisfied. The injection into the potential lead is two-staged (§5.5.1). A and A′ are adjustable in-phase and quadrature sources respectively. By two-staging, the necessary injection accuracy for comparatively large values of $\Delta \varepsilon / \varepsilon$ of up to 1000 parts per million is achievable.

The adjustable voltage sources are all driven from the current winding so that there is no need for the auxiliary isolation transformer T_A as in the previous section.

6.1.6 Equal power resistance bridge
This bridge was invented by Cutkosky to solve a problem which arises in the accurate comparison of resistors of different values, for example, of decade values of resistors used to set up a resistance scale. If the comparison is made on a n:1 voltage ratio bridge such as that described in §6.1.5 and shown in

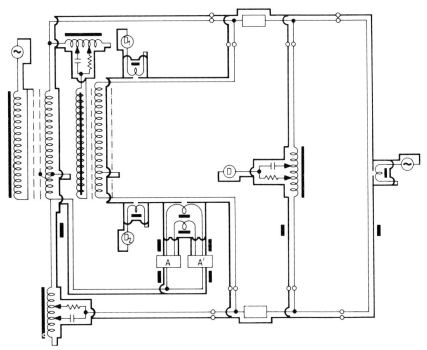

Figure 6.6 *A bridge for comparing four terminal-pair components, based on a 10: – 1 voltage ratio transformer.*

figure 6.7, the higher value resistor dissipates n times the power of the lower so that the dissipated power levels at which the value of the standards are defined (§2.2.1) decrease as n^p as p stages of scaling down are accomplished. This rapidly leads to an impossibly low power level at which the lower values of resistor should be measured. The alternative is to determine the load corrections of the resistors, but this also inevitably involves a measurement

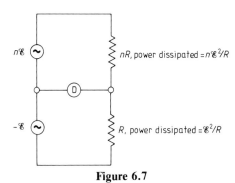

Figure 6.7

with a resistor dissipating little power, with the consequent problems of signal-to-noise ratio.

The same situation holds in reverse if a $n{:}1$ current ratio bridge is used in which the currents flowing through the two resistors when driven by the same voltage are compared with a current transformer as in figure 6.8, that is, the lower value resistor now dissipates n times the power of the higher.

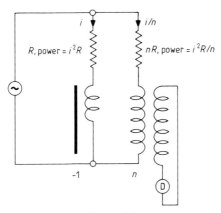

Figure 6.8

The sensitivity of these bridges is limited by the power W allowed to be dissipated in the resistor dissipating the most power. An isolated resistor of value R at a temperature T with a potential difference ε between its terminals, generates a noise EMF ε_n, with $W = \varepsilon^2/R$, and hence a measure of the sensitivity with which its value can be measured is

$$\varepsilon^2/\varepsilon_n^2 = W/4kT\delta f = S$$

where the detector bandwidth is δf and k is Boltzmann's constant. For the $10{:}-1$ voltage bridge the resolution is limited by the power dissipated in the larger valued resistor $10R$. The impedance presented to the detector terminals is $10R/11$, and so the sensitivity is

$$\varepsilon^2/\varepsilon_n^2 = \frac{(WR/10)}{[4kT\delta f(11/10R)]} = 0.11S.$$

In the case of the $10{:}-1$ current bridge, the resolution is limited by the current through the larger valued resistor, and we find

$$i^2/i_n^2 \simeq 0.09S.$$

This situation suggests the solution that if two resistors whose value are in the ratio $n^2{:}1$ are supplied with voltages in the ratio of $n{:}1$ and the consequent $1{:}n$ current ratio is measured, the resistors will dissipate equal

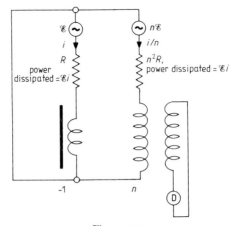

Figure 6.9

power. This is illustrated in figure 6.9. The signal-to-noise ratio of the last arrangement is also the most favourable.

For the equal power bridge, both resistors contribute equally to the noise current in the detector, and

$$i^2/i_n^2 = 0.5S.$$

The principle can be implemented by replacing the combining network between the components to be compared of the previous bridge with a $10: -1$ current transformer, and adding supplementary sources via injection transformers to fulfil the defining conditions $\varepsilon = 0$, $i = 0$, at the lower potential port of each component.

Figure 6.10 shows the principle of a network which we have used. The only auxiliary source which needs to be accurate is $-\Delta \varepsilon$ which adjusts the $10: -1$

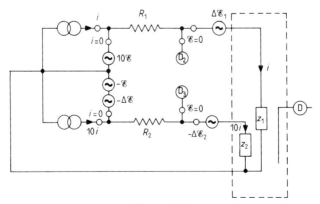

Figure 6.10

potential ratio. When the bridge is balanced, and the departure from a nominal $10: -1$ ratio of the current transformer has been accounted for by calibration, $\Delta\varepsilon$ determines the ratio of the impedances R_1 and R_2 to be compared.

The small potential drops in the internal impedances z_1 and z_2 (mainly arising from the resistance of the windings of the current transformer, with a little leakage inductance) have to be opposed to obtain nulls at the low potential ports of R_1 and R_2 with injected voltages $\Delta\varepsilon_1$ and $-\Delta\varepsilon_2$. The detector winding of the current transformer should be shorted when balancing detectors D_2 and D_3, as this ensures that they will be adjusted under the condition that will hold when the bridge is fully balanced, namely D balanced and zero flux in the current transformer core.

Then $-\Delta\varepsilon$ can be adjusted without substantially altering the previous balances to null the main detector D to obtain zero flux in the core of the current transformer.

The full circuit diagram of the bridge is drawn in figure 6.11. The adjustment of the $10i$ current source and the second core excitation is carried out, as in §6.1.5, to null D_5 and D_6. The role of the components can be

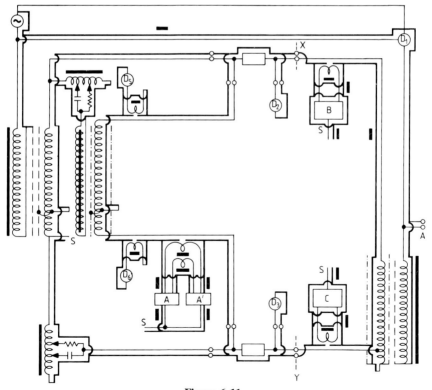

Figure 6.11

identified by comparison with figure 6.10; B and C are the means of
providing $\Delta\varepsilon_1$ and $-\Delta\varepsilon_2$, and the detector winding can be shorted at A.

The correction to the nominal ratio of the current transformer may be
determined by finding the apparent ratio of two equal capacitors, for which
purpose the connections at X and Y to the current transformer are
interchanged so that the network becomes a nominal $1:-1$ bridge. After
correction for the departure from its nominal ratio of the voltage
transformer, the mean of the two readings obtained before and after
interchange of the two capacitors is the desired correction to the nominal
current transformer ratio. If, however, the voltage dependence of the
capacitors is comparable with, or larger than, the uncertainty sought, a
correction will have to be made also for the difference in the values of the
capacitors when connected to the high and low potentials of the voltage
transformer.

6.2 Bridges to measure the ratio of unlike impedances_____

6.2.1 R–C. The quadrature bridge
This network, which has already been discussed theoretically (§2.6.3) was
devised principally for deriving the unit of resistance, the ohm, from that of
capacitance, the farad, realised by a calculable capacitor (§3.1).

It may also be used to establish resistance standards of small phase angle.
The principle of the method is first to use the quadrature bridge to adjust the
sum of the phase angles of two resistors to be zero, and then by substitution,
again using the quadrature bridge, to adjust their phase angles to be equal.
If, by an iterative procedure, these two conditions can be simultaneously
satisfied, each resistor must have zero phase angle. The phase angles of other
resistance standards can then be calibrated by using, for example, the
comparison bridges of §§6.1.2–6. It is assumed that the capacitances C_1 and
C_3 of figures 2.28 and 2.29 in §2.6.3 have either negligibly small phase angles
for the present purpose or that the sum of their phase angles

$$G_1/\omega C_1 + G_3/\omega C_3 = \Phi_C$$

(equation (2.19)) is known. We also assume that a third resistor R_5 similar to
the R_2 and R_4 is available. Denoting the phase angles of these latter
components by Φ_2, Φ_4 and Φ_5 we have, where $\delta\Phi$ corresponds to the
quadrature component required to balance the bridge,

$$\Phi_C + \delta\Phi_1 + \Phi_2 + \Phi_4 = 0 \tag{6.3}$$

and, by permuting two of the three resistors used,

$$\Phi_C + \delta\Phi_2 + \Phi_4 + \Phi_5 = 0 \tag{6.4}$$

$$\Phi_C + \delta\Phi_3 + \Phi_5 + \Phi_2 = 0 \tag{6.5}$$

hence

$$\Phi_4 = -(\Phi_C + \delta\Phi_1 + \delta\Phi_2 - \delta\Phi_3)/2$$

with similar expressions for Φ_2 and Φ_5.

If the resistors are provided with an adjustment (for example, an adjustable shunt capacitance to modify their phase angles is discussed in §3.3.6) they can be trimmed to have small phase angles. With R_2 and R_4 in use, and the bridge balanced, the bridge balance element (IVD or variable component) is altered by an amount which corresponds to the value $-\Phi_4$ and the phase angle trim on R_4 is adjusted to balance the bridge. Then it is again altered by an amount corresponding to $-\Phi_2$ and the trim on R_2 is adjusted to null the bridge again. R_2 is replaced by R_5, the quadrature balance altered by $-\Phi_5$ and the trim on R_5 adjusted to balance the bridge. Then, as a check, R_4 is replaced by R_2 when the amount the bridge is out of balance should be within the tolerance required on the trimming of the phase angles of $R_{2,4\,\mathrm{and}\,5}$. The procedure may be iterated if necessary.

6.2.2 The quadrature bridge—a two terminal-pair design

We describe a design originally used by Thompson. The principal components C_1, C_3 and R_2, R_4 are regarded as being defined in the two terminal-pair manner at the ends of leads which are therefore regarded as being a part of these components. The leads stay with the components when they are used in other bridges to determine the ohm from the calculable capacitor. The capacitors defined in this way cause no problems, but the definition of the resistors is limited by the varying contact resistance of the connectors terminating the cables. These contact resistances are not reproducible to better than $10^{-4}\,\Omega$, and as R_2 and R_4 are each $2 \times 10^4\,\Omega$, the uncertainty is limited to 1 or 2 in 10^8. Also, adjustable components are used rather than fixed components and adjustable voltages to make the auxiliary balances. Nevertheless this bridge is capable of an uncertainty approaching 1 in 10^8.

The network is drawn in figure 6.12 with the components labelled as in figure 2.29 of §2.6.3. The potential $-j\varepsilon$ at the junction of R_2 and C_3 is set up with the auxiliary adjustable capacitor fed from the -2ε tap and an adjustable resistor and associated phase trim capacitor from the $+2\varepsilon$ tap.

C_1, C_3, R_2 and R_4 are the components between which the relationship $1 + \alpha = \omega^2 C_1 C_3 R_2 R_4$ holds when the bridge is balanced (provided the product of the phase angles of R_2 and R_4 has been made sufficiently small by, for example, the adjustment procedure of §6.2.1). The detector combining network is completed by C_a, C_B, R_A and R_b and α is the difference from the mean of the in-phase readings obtained when the auxiliary capacitance bridge of 50 pF capacitors and detector D_2 is balanced, the connections to the rest of the bridge at X and Y are interchanged, and the capacitance bridge rebalanced. This mean represents equality of the output voltages at X and Y of the transformer and injection system. A suitable balance procedure for the quadrature bridge itself is to break the connection at X and Y temporarily and short the bridge side of each so that the nominal potential of $-j\mathscr{E}$ can be adjusted. This is done by altering the values of the components from the $\pm 2\mathscr{E}$ taps to null D_1. Then, after restoring the connection at X and Y, the '$-j\mathscr{E}$' balance is in turn temporarily upset by, for example, shorting the variable 5 nF capacitor and the detector network balanced by altering C_a and R_b and its trim capacitor. Finally, the short across the variable 5 nF is removed and α and β adjusted to again balance D_1. The whole balance procedure is then repeated a sufficient number of times that no readjustment of α and β is required to the full resolution desired. The accuracy with which

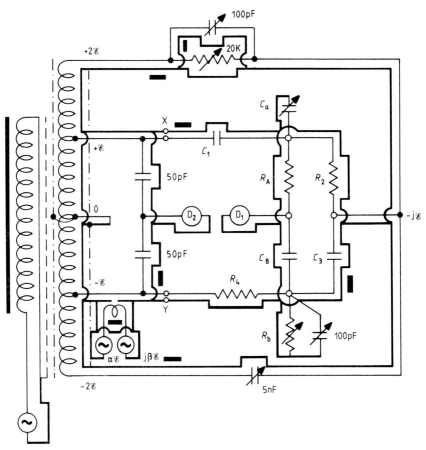

Figure 6.12 *A two terminal-pair quadrature bridge.* $C_1 = C_3 = 5\,\text{nF}$, $R_2 = R_4 = 20\,\text{k}\Omega$.

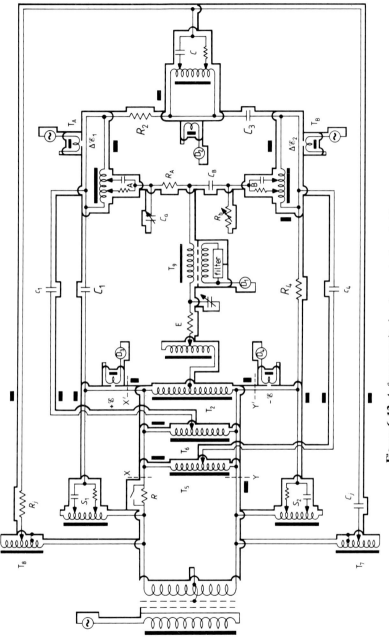

Figure 6.13 *A four terminal-pair quadrature bridge.*

the auxiliary balances have been made can be tested when the bridge is fully balanced by altering in turn, for example, the variable 5 nF capacitor and R_b by a fraction, 1 in 10^3 say, of their value. Since the main balance depends on only the product of these small quantities, it should not change by more than $10^{-3} \times$ (setting accuracy of the other).

6.2.3 The quadrature bridge—a four terminal-pair design

Improved accuracy of definition of the standards can be obtained by using a four terminal-pair configuration at the expense of increased complexity in the bridge design. A circuit has been published by Cutkosky (1970) and the version which the authors have used is shown in figure 6.13.

The four terminal-pair components are C_1, C_3, R_2 and R_4. They are all joined in series in a way equivalent to realising their defining conditions by using combining networks (§§2.7 and 5.5.4) A, B and C. Networks A and B are adjusted in turn so that the injected voltage sources $\Delta \varepsilon_1$ and $\Delta \varepsilon_2$ have no effect on the main detector D_1, thus producing a condition equivalent to $\Delta \varepsilon_1$ and $\Delta \varepsilon_2$ being zero. Network C is adjusted so that a detection transformer and detector D_2 connected between the upper potential ports of R_2 and C_3 is nulled, thus fulfilling the condition $i = 0$ at these ports. As in the previous section, C_a, R_A and R_b, C_B form a combining network which, when adjusted, causes the detector D to be immune to departures of the auxiliary source created by the components $T_{7,8}$, R_j and C_j from the required value $-j\varepsilon$. Only one source is required, to replace in turn the shorts in the inputs of T_1, T_A and T_B and similarly only one detector to replace the shorts in detector positions D_1–D_4 respectively, in turn to make the various balances as needed. A and B are adjusted by removing the main bridge source from the supply and isolation transformer T_1, replacing it with a short, plugging the supply in turn into T_A and T_B and adjusting A and B respectively for a null on the main detector D_1.

The source $-j\varepsilon$ applied to the upper current ports (§2.2.8) of R_2 and C_3 is provided by the adjustable IVDS T_7 and T_8 in conjunction with the components C_j and R_j and the combining network C which is adjusted to null D_2. The main detector combining network is adjusted by altering C_a and R_b to null the main detector D_1 when T_7 and/or T_8 are temporarily maladjusted, and similarly T_7 and T_8 are adjusted to null D_1 when C_a and R_b are temporarily maladjusted. The main balance adjustment to null D_1 when T_7, T_8 and C_a and R_b have been adjusted is made with IVDS T_5 and T_6 which, in conjunction with c_1 and c_4 have the effect of putting small variable capacitances in parallel with C_1 and R_4. The defined potentials $\pm \varepsilon$ are applied to the upper potential ports of C_1 and R_4 by the $1:-1$ transformer T_2, the condition $i = 0$ being tested by the defining transformers and associated detectors D_3 and D_4 which are nulled by the adjustable current sources, S_1 and S_2.

The departure of T_2 from a perfect $1:-1$ ratio is eliminated by reversing the inputs to it at X and Y and the connections to D_3 and D_4 at X′ and Y′.

This also reverses the sense of T_5 and T_6 so that their readings corresponding to zero voltage applied to c_1 and c_4 need not be determined. Because there may exist a small potential difference between the centre tap of T_2 and its outer at its exit port, a combining network E here is adjusted so that when this potential is temporarily increased by switching a resistor R of a few ohms in series with one input of T_2, the reading of the main detector at D_1 does not change.

All these adjustments are iterated as necessary, but the convergence is very rapid. The main detector D_1 is coupled into the bridge with the matching transformer T_9 and unwanted harmonics are eliminated with the filter described in §5.1.1.

There are fifteen meshes in the network shown, and each is provided with an equaliser whose effect on the total network is evaluated by the method described in §4.4.7.

The EMF applied to c_1 is $(0.5 - k_1)2\varepsilon = (1 - 2k_1)\varepsilon$ and hence the admittance of C_1 is augmented by $c_1(1 - 2k_1)$, where k_1 is the setting of IVD T_6, so that the bridge balance condition of §2.6.3, equation (2.15) becomes

$$\omega^2[C_1 + c_1(1 - 2k_1)]R_2C_3R_4 = 1.$$

On reversing T_2, T_5 and T_6, the condition for balance at a reading of T_6 of k_1 is

$$\omega^2[C_1 + c_1(2k_1' - 1)]R_2C_3R_4 = 1$$

and taking the mean of these two results, which eliminates any small departure from a nominal $1 : -1$ ratio of T_2, we have finally for the in-phase balance condition

$$\omega^2[C_1 + 2(k_1' - k_1)c_1]R_2C_3R_4 = 1.$$

For the quadrature adjustment made by T_5, the phase defect represented by a shunt capacitance c_4 of R_4, is similarly modified by $2(k_2 - k_2')c_4$ so that the relation of equation (2.19) between the phase defects becomes

$$G_1/\omega C_1 + R_2\omega C_2 + G_3/\omega C_3 + R_4\omega[C_4 + 2(k_2 - k_2')c_4] = 0.$$

The bridge has been constructed with components whose nominal values are $C_1 = C_3 = 1\,\text{nF}$, $R_2 = R_4 = 10^5\,\Omega$, $c_1 = c_4 = 1\,\text{pF}$. The nominal working frequency was $\omega = 10^4\,\text{rad s}^{-1}$.

With $\varepsilon = 10\,\text{V}$, giving $1\,\text{mW}$ power dissipation in the resistors, and a detector time-constant of 3 s, a resolution and accuracy approaching 1 in 10^9 is achieved.

6.2.4 *C–L. The parallel tuned circuit approach*

The value of a self-inductor of value L can be found by resonating it in a parallel tuned circuit at a measured angular frequency ω with an adjustable capacitor of value C so that the impedance of the combination is resistive. Both the conductance G and the residual capacitance δC (if exact resonance is not achieved) can then be measured on a suitable bridge, as for example the two terminal-pair IVD bridge described in §6.1.2; the bridge can then separately measure the value of C and its shunt conductance G_c at frequency ω (see figure 6.14). Account must be taken of the small inductance δL and resistance δR of the leads needed to connect the inductor to the capacitor. Remembering that the series combination of $L + \delta L$, $R + \delta R$ presented to the terminals of the capacitor can also be represented by the parallel combination L', G', as shown in figure 6.15, we have the following relationships

$$G' = G - G_c \tag{6.6}$$

$$L' = \omega^2(C - \delta C)^{-1} \tag{6.7}$$

$$D = \omega L' G' \tag{6.8}$$

$$L = L'(1 + D^2)^{-1} - \delta L \tag{6.9}$$

$$R = \omega L D - \delta R. \tag{6.10}$$

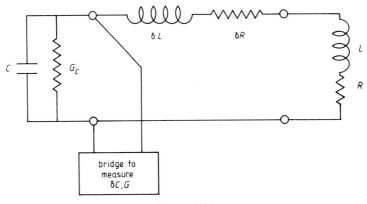

Figure 6.14

The method is most suitable for inductors of small D (high Q); D is otherwise difficult to measure with sufficient accuracy for its uncertainty not to contribute substantially to the uncertainty of the series inductance L found from equation (6.9). This situation is not really a limitation because the method can be used to establish the value of a special high Q inductor in terms of which other inductance values can be found by an inductance comparison bridge such as that of §6.1.2.

The principal problems of this simple technique are caused by the ill-defined nature of the open coil short solenoid inductor likely to be used to give a high Q of the order of 100 at 1 kHz (§3.5.2). The electric and magnetic fields of the inductor extend appreciably into the space surrounding it and will be modified by nearby conducting, dielectric or magnetic material. The

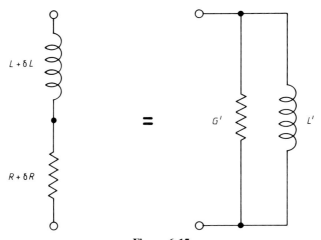

Figure 6.15

parameters L and R are the lumped impedances which represent the actual impedance presented to the measurement circuit at the terminals of the inductor. They therefore include the effect of capacitance and mutual inductance to the surroundings of the inductor which include the screened leads needed to connect the inductor terminals to the rest of the apparatus. These difficulties may be sufficiently well overcome by regarding the inductance as being defined when used with certain leads in a definite configuration near the terminals, and to retain these leads for use in any bridge which relates the value of an inductor calibrated in this way to that of other inductors. Care is still needed, however, to ensure that either the relative potentials of the inductor windings and any screens on these leads are similar in the two situations, or that any such capacitative effects are negligible.

The adaptor described in §5.4.1 provides a suitable lead configuration for connection between the two terminal defined inductor and a two terminal-pair defined capacitor and bridge.

The magnitudes of the quantities involved in a measurement that the authors carried out in association with Mr M J Swan were

Frequency	1 kHz
5 values of L of	0.05 to 1 H
needing values of C of	0.5 to 0.025 μF
values of D ($= 1/Q$)	0.1 to 0.01
uncertainty of determinations	4 to 10 parts per million.

It should be remembered that this is a frequency-dependent measurement so that attention to the defects of detectors is necessary regarding insufficient selectivity and linearity with consequent intermodulation distortion (§5.1.1).

6.2.5 C–M

Because of the general applicability and accuracy of ivds and ratio transformers, mutual inductors, whether fixed or of variable value, are now much less used than formerly. It is therefore chiefly for the sake of completeness that we discuss a suitable network for calibrating mutual inductance in terms of capacitance. The network of figure 6.16 does however serve to illustrate the universal applicability of four terminal-pair defined components in a conductor-pair network.

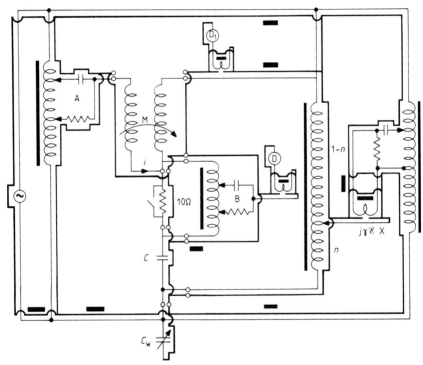

Figure 6.16 *A bridge for comparing the values of a mutual inductor and a capacitor.*

Most existing mutual inductors are ill defined in the sense that they are not constructed to have zero magnetic field external to their case; indeed their external electric field is rarely zero either. Moreover, they are usually provided with single terminal screw binding-posts rather than coaxial terminations. These limitations mean that they must be sited at a distance equal to many times a typical dimension from any conducting material to avoid eddy current interactions, and especially from ferromagnetic material which would modify their external magnetic field. The potential of one part of an inductor with respect to another part is defined, but not of the inductor as a whole with respect to its surroundings. Mutual inductors also present special problems because they have internal impedance associated with the potential, or secondary terminals which is comparable with that of the mutual inductance. For these reasons, the attainable accuracy is inherently limited to perhaps 1 part per million or worse. Fortunately there seems to be no need for more accurate values of mutual inductance.

In comparing a mutual inductance with a capacitance, because a mutual inductor has a much smaller phase defect than a self-inductor, the limitations imposed by a low Q in the latter do not arise.

We describe a simple bridge based on the four terminal-pair IVD bridge of §6.1.3 which enables the unit of mutual inductance to be derived from that of capacitance to modest accuracy at a suitable impedance level (for example, 10 mH and 1 μF and frequency $= 10^4$ rad s^{-1}). At this level, the capacitance of the mutual inductor to the surroundings is likely to be about 10 pF, and the impedance of this is large compared with the mutual impedance of the device. Note that the measurement is to be made with a frequency-dependent bridge, and the strictures of §5.1.1 must be borne in mind.

The principle of the comparison is simple; a current i flows through the primary of M and through C in series; the potential differences $j\omega Mi$ across the secondary of M and $-ji/\omega C$ across the potential ports of C are compared with the ratio of potentials provided by an IVD and the unwanted potential drop along the current lead between the potential terminals of M and C is eliminated with a combining network.

M must be connected to give the correct sign of EMF to balance the bridge. Both the inductor and the capacitor C are defined as four terminal-pair components; the inductor can be interfaced for this purpose by using the adaptor of §5.4.1, the screened twisted leads being made long enough so that the metal of the connectors is far enough from the inductor that it does not perturb its value significantly.

The balance procedure is as follows.

(i) With the 10 Ω resistor shorted and X open-circuited, D_1 is nulled by adjusting the current source A.

(ii) With connection restored at X, the ratio $\varrho = n/(1-n)$ of the main IVD is adjusted to null D.

(iii) (i) and (ii) are iterated until convergence to a reasonable accuracy is obtained.

(iv) With the short on the 10 Ω resistor removed, the combining network B is adjusted to null D.

(v) The 10 Ω resistor is shorted again, and ϱ adjusted again to null D. ((iv) and (v) are iterated as necessary.)

(vi) The Wagner component C_W is adjusted by nulling the detector which temporarily replaces the short at X for this purpose.

(vii) The whole procedure thus far is iterated as necessary until convergence is obtained to the desired accuracy.

The equations governing the balanced bridge, using the series lumped impedance representation of M (figure 3.31) and C (figure 3.4(c)) are

$$(j\omega M + r)/(-j/\omega C' + r_C) = (1 - n - j\gamma)/(n + j\gamma).$$

Whence, equating the real and imaginary parts,

$$-\omega\gamma M + rn = -\gamma/\omega C' + r_C(1 - n) \qquad (6.11)$$

$$\omega Mn + r\gamma = -(1 - n)/\omega C' - r_C\gamma. \qquad (6.12)$$

Since, for standards of small phase defect

$$r \ll \omega M \qquad\qquad r_C \ll 1/\omega C' \qquad\qquad \gamma \ll n \text{ or } 1 - n$$

from equation (6.12) we have the approximate expression

$$M = -(1 - n)/n\omega^2 C' \qquad (6.13)$$

and substituting this in equation (6.11),

$$r = r_C(1 - n)/n - \gamma/\omega C' n^2. \qquad (6.14)$$

It should be remembered that the effective value of a mutual inductor changes with frequency because of the inductance of the windings combining with the interwinding capacitances, and the latter depend on which ends of the windings are brought together at the common point. Usually the series aiding connection has a larger change with frequency than the series opposing one. The circuit described can only be balanced for the series opposing connection.

6.3 Bridges posing or solving particular problems_____

6.3.1 M–M IVD bridge

In this section a bridge is described which compares two mutual inductors, and a balance can only be obtained if the mutual inductors are of opposite sign.

In figure 6.17 the mutual inductances M_1 and M_2 are treated as four terminal-pair components; devices of older construction having exposed single terminals must be converted using the adaptor described in §5.4.1. The current sources A and B feeding the primary windings of M_1 and M_2 must be adjusted both to null D_2 (thus ensuring that the same current traverses both primary windings) and to null D_1 so that when D_m is nulled by adjusting the main IVD on the right and the quadrature source on the extreme right, no current is drawn from either secondary winding. The ratio

main IVD   quadrature source

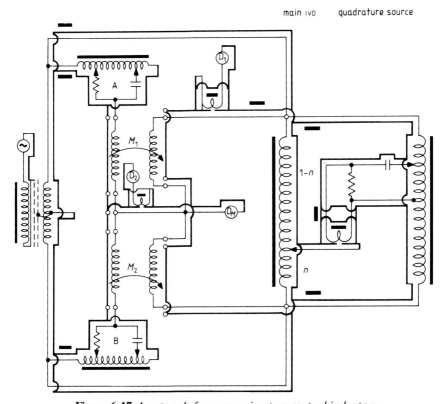

Figure 6.17 *A network for comparing two mutual inductors.*

$M_1/M_2 = (1-n)/n$; the phase defect difference is derived from the fractional amount of injected quadrature voltage.

A full description of a slightly more elaborate bridge, and much useful information on the construction and electrical definition of small-valued mutual inductors has been given by Homan (1966).

6.3.2 A bridge for use with a calculable capacitor

In this section we describe a bridge used with a Thompson–Lampard calculable capacitor (§3.1) to calibrate a fixed 10 pF capacitor in terms of the SI capacitance unit realised by the calculable capacitor.

The bridge must resolve the capacitances $C_{1,2}$ of the calculable capacitor (which lie in the range 0.2–0.7 pF), usually at a frequency of about 1592 Hz, to a resolution of 5×10^{-9} pF in order to calibrate the 10 pF capacitor C_s to an accuracy of 1 in 10^8. The ratios near to nominal values of $C_{1,2}$ equal to 0.2, 0.3, . . . , 0.7 pF need to be calibrated to this accuracy. In the method we describe, use is made of the calibration of intermediate taps of a 10: −1 voltage ratio transformer obtained by the calibration technique of §4.4.2.

The bridge can also be calibrated by adding successive numbers of 10 pF capacitors in parallel on the high potential side whilst balancing against a 1000 pF capacitor on the low potential side. The values of the 10 pF capacitors have to be individually related to that of the 1000 pF capacitor, for example by using the bridge described in §6.1.5 in two successive decade steps with a stable 100 pF capacitor as an intermediate value.

The bridge is as represented in figure 6.18. Since only the small admittances of $C_{1,2}$ and C_s are to be compared, the simple two terminal-pair definition of them is adequate. The bridge is based on the $10: -1$ ratio of a voltage transformer, but the ability to encompass any capacitance value within the above range is conferred by adding an eight decade IVD between the 0 and -1 taps of the transformer. Further resolution is obtained by interpolation of the last decade. If the IVD ratio is nominally k, we then have

$$10(1 + \alpha_{10})C_{1,2} = (1 + \alpha_{-1})(k + \delta k)C_s$$

where α_{10} is the departure from nominal of the 10 tap voltage; α_{-1} that of the -1 tap; and δk is the departure from nominal of k.

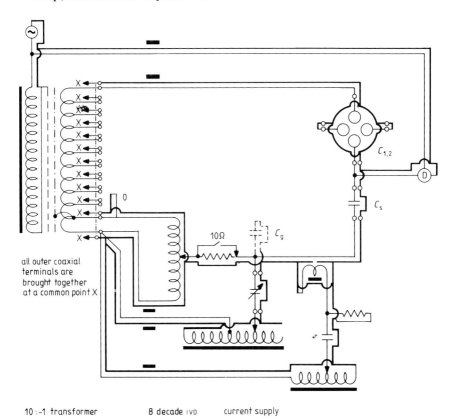

all outer coaxial
terminals are
brought together
at a common point X

10 :-1 transformer   8 decade IVD   current supply

Figure 6.18

Because multi-decade IVDs have output impedances whose greatest values are at least a factor of ten greater than those of a well designed 10: − 1 transformer it is important that the eight decade IVD used in this network is operated in such a way that its output is not required to deliver current. This is accomplished by the mesh consisting of the auxiliary IVD connected into the − 1 transformer port feeding a variable capacitor which in turn is connected into the output from the eight decade IVD. It is adjusted to be equal in value to C_g which represents the total capacitive load of cables, the input admittance of C_s and of C_s itself when the setting of the auxiliary IVD is equal to that of the first three or four decades of the eight dial IVD. This adjustment is accomplished simply by putting a temporary 10 Ω resistor in series with the output of the eight dial IVD and adjusting the variable capacitor until the main bridge balance is restored. Shorting the 10 Ω resistor as shown should not then result in a change in the main bridge balance, so that a condition has been achieved whereby the potential drop down the 10 Ω resistor has been made negligible, that is, there is negligible current through it drawn from the eight decade IVD's output. Since the auxiliary IVD, configured as shown, provides twice the voltage at its output as the eight dial IVD if the two IVDs have the same dial settings, C_g and the variable capacitor form a capacitance bridge with shorted 10 Ω resistor in place of a detector. The bridge current through C_s and $C_{1,2}$ then is entirely provided by the variable capacitor and the auxiliary IVD which form a load, varying with the IVD settings, on the − 1 tap of the transformer. The slight effect of this load, as well as that of the lead and input capacitance of $C_{1,2}$ on the 10 tap and of the auxiliary and quadrature balance IVDs must be accounted for by investigating to what extent they affect the bridge balance. This can be done by adding more capacitative loading in a known ratio to the actual loads to the − 1 and 10 taps.

The bridge may be calibrated at balance points corresponding to 0.1 pF increments of $C_{1,2}$ by connecting a stable 10 pF capacitor in place of $C_{1,2}$ and a stable 100 pF capacitor in place of C_s. The ratio of values of these two capacitors is first determined by balancing the bridge with the eight decade IVD at a ratio of near unity. Then balances are obtained with the 10 pF capacitor connected into successively lower taps. The corrections to IVD settings in the vicinity of these balance positions can be determined from the corrected voltages of the 10: − 1 voltage transformer taps once the loading presented to these taps and the change in apparent value of the 10 pF and 100 pF capacitors with applied voltage have been accounted for.

Finally, we may note that the linearity of a Thompson–Lampard calculable capacitor, that is, whether successive sections along the capacitor electrodes yield the same value of C_s, can readily be tested by connecting into the detector junction point another stable 10 pF capacitor supplied via an IVD from the − 1 transformer tap. This arrangement shunts C_s with a capacitance whose effective value is 10 times (IVD setting) pF, and the working part of the capacitor $C_{1,2}$ can be selected by altering the apparent value of this shunting capacitor.

References

Corney A C 1979 A universal four-pair impedance bridge *IEEE Trans. Instrum. Meas.* **28** 211–15

Cutkosky R D 1970 Techniques for comparing four terminal-pair admittance standards *J. Res. NBS* **74D** 63–78

Hanke R 1978 Precise Kelvin double bridge for measuring dissipation factors and capacitances up to 1F *IEEE Trans. Instrum. Meas.* **27** 434–6

Homan D N 1966 Some techniques for measuring small mutual inductances *J. Res. NBS* **70c** 221–6

Igarashi T, Koizumi Y and Kanno M 1968 Determination of an absolute capacitance by a horizontal cross-capacitor *IEEE Trans. Instrum. Meas.* **17** 226–31

Kibble B P, Rayner G H and Swan M J 1980 On obtaining the Henry from the Farad *National Physical Laboratory Report DES 63*

Thompson A M 1975 Standards for the measurements of phase angles *Proc. IEE* **122** 576–8

Zapf T L 1961 Calibration of inductance standards in the Maxwell–Wien bridge circuit *J. Res. NBS* **65c** 183–8

Index